Global Migration Issues

Volume 6

Series editor
Frank Laczko

This book series contributes to the global discussion about the future of migration policy through the publication of a series of books on emerging migration issues. Most reports on migration policy tend to focus on national or regional perspectives; books in this series will focus on global policy challenges, such as the impact of climate change or the global economic crisis, on migration.

This series is closely linked to the production of IOM's *World Migration Report*. Some of the books in this series will be based on research which has been prepared for the *World Migration Report*.

The series also includes a special focus on the linkages between migration and development, and the themes discussed each year at the Global Forum on Migration and Development (GFMD), given the growing policy interest in harnessing the benefits of migration for development.

More information about this series at http://www.springer.com/series/8837

Andrea Milan • Benjamin Schraven
Koko Warner • Noemi Cascone
Editors

Migration, Risk Management and Climate Change: Evidence and Policy Responses

Editors
Andrea Milan
Institute for Environmental & Human
   Security
United Nations University
Bonn, Germany

Koko Warner
Institute for Environmental & Human
   Security
United Nations University
Bonn, Germany

Benjamin Schraven
German Development Institute
Bonn, Germany

Noemi Cascone
Institute for Environmental & Human
   Security
United Nations University
Bonn, Germany

At the time of publication, Andrea Milan, Koko Warner and Noemi Cascone are not affiliated with UNU-EHS anymore.

ISSN 2213-2511   ISSN 2213-252X (electronic)
Global Migration Issues
ISBN 978-3-319-82694-3   ISBN 978-3-319-42922-9 (eBook)
DOI 10.1007/978-3-319-42922-9

© Springer International Publishing Switzerland 2016
Softcover reprint of the hardcover 1st edition 2016
This work is subject to copyright. All rights are reserved by the Publisher, whether the whole or part of the material is concerned, specifically the rights of translation, reprinting, reuse of illustrations, recitation, broadcasting, reproduction on microfilms or in any other physical way, and transmission or information storage and retrieval, electronic adaptation, computer software, or by similar or dissimilar methodology now known or hereafter developed.
The use of general descriptive names, registered names, trademarks, service marks, etc. in this publication does not imply, even in the absence of a specific statement, that such names are exempt from the relevant protective laws and regulations and therefore free for general use.
The publisher, the authors and the editors are safe to assume that the advice and information in this book are believed to be true and accurate at the date of publication. Neither the publisher nor the authors or the editors give a warranty, express or implied, with respect to the material contained herein or for any errors or omissions that may have been made.

Printed on acid-free paper

This Springer imprint is published by Springer Nature
The registered company is Springer International Publishing AG Switzerland

Contents

Part I Mountain Areas

1 **An Index Based Assessment of Vulnerability to Floods in the Upper Indus Sub-Basin: What Role for Remittances?** 3
Soumyadeep Banerjee, Muhammad Zubair Anwar,
Giovanna Gioli, Suman Bisht, Saleem Abid, Nusrat Habib,
Sanjay Sharma, Sabarnee Tuladhar, and Azeem Khan

2 **Role of Remittances in Building Farm Assets in the Flood Affected Households in Koshi Sub-Basin in Nepal** 25
Soumyadeep Banerjee, Bandita Sijapati, Meena Poudel,
Suman Bisht, and Dominic Kniveton

3 **Migration as a Risk Management Strategy in the Context of Climate Change: Evidence from the Bolivian Andes** 43
Regine Brandt, Raoul Kaenzig, and Susanne Lachmuth

4 **Circular Migration and Local Adaptation in the Mountainous Community of Las Palomas (Mexico)** 63
Noemi Cascone, Ana Elisa Peña del Valle Isla, and Andrea Milan

Part II Low-Lying Areas

5 **Household Adaptation Strategies to Climate Extremes Impacts and Population Dynamics: Case Study from the Czech Republic** 87
Robert Stojanov, Barbora Duží, Ilan Kelman, Daniel Němec,
and David Procházka

6 **Moving Beyond the Focus on Environmental Migration Towards Recognizing the Normality of Translocal Lives: Insights from Bangladesh** 105
Benjamin Etzold and Bishawjit Mallick

Part III Small Islands

7 Good Fishing in Rising Seas: Kandholhudhoo, Dhuvaafaru, and the Need for a Development-Based Migration Policy in the Maldives 131
Andrea C. Simonelli

8 The Reason Land Matters: Relocation as Adaptation to Climate Change in Fiji Islands 149
Dalila Gharbaoui and Julia Blocher

9 The Role of Remittances in Risk Management and Resilience in Tuvalu: Evidence and Potential Policy Responses 175
Sophia Kagan

Part IV Policy

10 Remittances for Adaptation: An 'Alternative Source' of International Climate Finance? 195
Barbara Bendandi and Pieter Pauw

11 Conclusion: Migration as Adaptation: Conceptual Origins, Recent Developments, and Future Directions 213
Robert McLeman

Introduction – Climate Change and Human Mobility After Paris

Introduction

Following the climate negotiations in Paris which made more space than ever for the issue of human mobility, against emerging scientific evidence, and against the background of an ongoing refugee crisis in Europe, it is time for the international community to pursue an evidence and needs-based protection framework for environmental migrants and people displaced by climate stressors.

The current refugee crisis in Europe is about a brutal civil conflict in Syria and not about climate change. However, it sends a signal about the kinds of human movements we will see in the future as climatic stressors, such as storms, droughts, heat waves and sea level rise increasingly impacts jobs, food security, and the stability of urban and rural areas.

Science

What does current science say about human mobility and climate change? Science points to widespread current and future biophysical impacts of anthropogenic climate change (IPCC 2012, 2014; Fung et al. 2010). Human mobility—migration, displacement, potentially planned relocation—are themes woven through the Fifth Report of the Intergovernmental Panel on Climate Change (IPCC 2014). The report notes emerging risks and threats that affect livelihoods, food security, and safety.

These key risks have political importance as well, because they inform the evaluation of "dangerous anthropogenic interference with the atmosphere" as laid out in Article 2 of the UN Framework Convention on Climate Change (UNFCCC). Article 2 outlines its ultimate objective as the 'stabilization of greenhouse gas concentrations in the atmosphere at a level that would prevent dangerous anthropogenic interference with the climate system… in order to allow ecosystems to adapt naturally to climate change, to ensure that food production is not threatened and to enable economic development to proceed in a sustainable manner'. One way to think about Article 2 is maintaining a 'safe operating space for humanity' (Röckstrom et al. 2009). These key risks are "potentially severe adverse consequences for humans and socio-ecological systems resulting from the interaction of hazards linked to climate change and the vulnerability of exposed societies and system" (IPCC 2014: 1043). These key risks include factors that have been directly linked to mobility pressures, in particular risks of food insecurity and breakdown of food systems, and risk of loss of rural livelihoods linked to insufficient water and reduced agricultural productivity.

Community-based empirical research across the world, such as the chapters in this book, indicate that people will move away from regions that climate change slowly renders uninhabitable, such as small island states in the Pacific affected by sea level rise and parts of South East Asia dealing with coastal erosion. They will move towards areas they hope will provide safe and sustainable livelihoods. Almost half of the world's population depends on agricultural production for their livelihoods, and this sector is most severely impacted by a changing climate. Evidence from this literature, including scholarly leaders publishing chapters in this edited volume, underscores that vulnerable households use different forms of human mobility to manage climatic risks. Climate impacts such as changes in rainfall variability (untimely rain, unseasonal and unexpected precipitation, or shortfalls in rain) affect the stability of household livelihoods, which in turn can negatively affect household income and consumption (Afifi et al. 2015; Warner and Afifi 2014). Pressures to move involve multiple interacting systems and stresses—crop production, prices, and increased food insecurity (Adger et al. 2014; Oppenheimer et al. 2014). Now and in the future, research suggests that indirect, transboundary, and long-distance impacts (Oppenheimer et al. 2014) are expected to drive human migration and displacement when thresholds for livelihoods, food security, and safety are breached (Klein et al. 2014).

Policy

All countries and governments will be affected by people on the move whether those countries are areas of origin, transit, or destination. People will move either in anticipation of climate stressors or in response to them.

Over the past decade, discussions about climate change and migration, displacement, and planned relocation have moved from limited research or policy discussion to growing, robust evidence and significant policy milestones. The international community needs a robust legal framework to guide efforts to assist people on the move because of climate stressors. At the current stage, people leaving their countries due to climate stressors are not considered refugees under the Geneva Convention, which specifies that a refugee is fleeing from a "well-founded fear of being persecuted for reasons of race, religion, nationality, membership of a particular social group or political opinion."

Climatic or environmental factors are not recognized as persecuting factors, and only some countries grant temporary protection status and waive visa requirements for migrants whose home country faces a severe natural hazard.

People who move are often particularly vulnerable and need a scope of assistance – from legal protection to access to labor markets, valid identification, and integration opportunities. The current situation in Europe shows us that we are not yet prepared for such large movements of people within or across borders. While the challenges of the current refugee situation are immense, governments and the international community are guided by national and regional policies that follow the clear normative and legislative framework of the Geneva Convention. Such policies and frameworks are currently not in place for environmental migrants.

In the autumn of 2015, over 100 governments have endorsed the Nansen Initiative Agenda for people displaced across borders in the context of natural disasters and the effects of climate change. This agenda, supported by the United Nations University as an Advisory Group member, helps point the way for filling legal gaps and providing evidence-based policy and operational support for vulnerable affected people. In Paris, the Conference of the Parties established a task force on human mobility to develop recommendations for the Warsaw International Mechanism for Loss and Damage (paragraph 50).

In the post-2015 world, all current signals point to a need to invest in research and policy analysis to develop a reference point which will not only help to protect vulnerable people but will also serve long-term sustainable development.

Chapters in This Volume

The largely evidence-based, case study-based chapters in this book reflect a collection of scholarly work that recognizes human migration as one of a number of attempts of vulnerable households to manage risks including climatic stressors. Chapters span three major veins of examination in the dynamics of migration linked to climatic stressors: the role of remittances in enhancing (or not) adaptive capacity of families that do have one or more migrant members; the interactions of decisions about livelihood security and how migration fits into those decision patterns; and the role of land tenure and related policies in migration and relocation in land-constrained areas like the Pacific ocean.

Remittances

In their chapter on the role of remittances, Banerjee et al. ask whether and how remittances help reduce the vulnerability of recipient households to a flooding in the Upper Indus subbasin. The vulnerability assessments find that remittance-recipient households are marginally less vulnerable than non-recipient households and are less likely to reduce food consumption during floods. Interestingly, for farming households, non-remittance-recipient households may demonstrate other forms of adaptive capacity, such as changing agricultural practices in response to floods.

In a second coauthored chapter, Banerjee et al. explore the relationships between mobility, remittances, and adaptive capacity in the rural Sagarmatha transect of Koshi subbasin of Nepal through building farm assets such as farm size, livestock, irrigation, and farm mechanization. They find little difference in the flood response strategy of remittance-recipient and non-recipient households but find that the longer a time a household receives remittances, the more likely it is to reduce its farm holding.

Kagan uses the lens of a case study in Tuvalu to analyze empirical evidence of the relationship between remittances and disaster risk management. The author finds that while remittances form a key part of coping strategies after a disaster, there is insufficient evidence to suggest that remittances improve ex ante risk management.

Bendandi and Pauw ask whether remittances could constitute international adaptation finance. They find that incentives for diaspora communities need to be provided in order to channel remittances toward adaptation. They conclude that remittances can help to support adaptation at household and community level.

Livelihoods

Brandt et al. analyze risk management and migration decisions in the face of climate variability and water scarcity in two rural areas near La Paz, Bolivia. Their findings correlate with that of the growing literature that social, economic, and environmental factors drive decisions about managing livelihood risks with migration.

Cascone et al. explore the potential of resilience-building measures and circular migration programs as part of household strategies to diversify livelihoods and manage risks associated with environmental and climate change in Las Palomas, Central Mexico. The authors find that sending one or more migrants abroad as a risk management strategy at the household level can allow the rest of the household to stay where they are and to increase their adaptive capacity through increased income and livelihood risks reduction.

Stojanov et al. examine household adaptation strategies in the face of floods between 1997 and 2012 in selected rural municipalities in the Bečva river basin in the northeastern part of the Czech Republic. Their research revealed a link between difficulty migrating and social consequences, meaning that the increasing occurrence of floods is a serious problem for residents who cannot leave, because they had limited opportunities for resettlement.

Etzold and Mallick find that translocal households with migrants employ livelihood choices, human rights, and freedoms that enhance their resilience to environmental and socioeconomic risks. They argue that it is necessary to move beyond framing migration as a failure of adaptation to environmental risks and instead recognize the normality of people's mobility, the persistence of regional migration systems, and the significance of the practices and structures that enable Bangladeshis to live secure translocal lives. Such a change in perspective has significant repercussions for the politics of climate change adaption and the management of migration.

Land

Simonelli examines migration and limits to adaptive capacity in the isolated Kandholhudhoo fishing community in the Maldives. The author proposes that policy responses are needed—particularly tailored to the vulnerabilities of small island states—which more fully utilize options for internal migration, with implications for population densities, island structural integrity, and economic resource bases.

Gharbaoui and Blocher examine the role of customary land tenure and land use in complex relocation processes in the Pacific. Against a historical analysis of ancestral and recent community relocation and land tenure in Fiji, the authors argue for participatory adaptive relocation processes which consult, cooperate, and negotiate with customary leaders of both sending and receiving communities at an early stage.

Finally, in the last chapter of the book, McLeman traces how migration in policy and research has increasingly been framed in terms of vulnerability and adaptation. The author examines critiques of this conceptualization and suggests promising avenues for further theoretical development and policy discussion.

Looking Forward

Looking forward, science is needed that will inform decisions about climate-resilient development pathways which includes human mobility. It is common for debates to form around normative questions such as whether different forms of mobility are a "positive" form of adaptation or an indicator of the severity of climate impacts. What will be important moving forward, however, is a focus on leaving no group of vulnerable people behind in the quest for improved human welfare. Climate change poses significant challenges to this overarching aim of the Sustainable Development Goals. What will emerge in the next rounds of research will be an understanding of human mobility as a global process of societies adjusting culturally, geographically, politically, and economically to the adverse effects of climate change.

Both the emerging science (IPCC 2014) and the Paris outcomes acknowledge the relationships between a range of climatic stressors and forms of human mobility, the need for actions that reduce vulnerability factors and enhance resilience factors for affected people, and principles that can guide support and work on climate-related human mobility. These major science and policy milestones in 2014 and 2015 thus provide insights into directions for research, policy, and operations in coming years:

- First, research can help fill gaps in understanding on factors which affect vulnerability or resilience of people who are moving and the networks they are part of (families, communities), and it offers insights into the factors and thresholds relevant to household decisions to move or not.
- Second, policies drawing on this research are needed to guide risk averting and minimizing actions, as well as actions to address human mobility related to climate change (displacement in particular).
- Third, action and support to address human mobility in the context of climate change will be needed which include participation of affected people, guided by the best available science and other knowledge systems (traditional, indigenous, local) and aimed at integrating these actions into relevant socioeconomic and environmental policies and actions.

Human mobility in the face of climate change is a risk management strategy and livelihood diversification strategy in the face of many pressures and aspirations to better human welfare. The chapters in this book examine evidence from Pakistan, Nepal, Bangladesh, Tuvalu, Bolivia, Mexico, the Czech Republic, and Pacific Region and bring cutting edge analysis, insights, and suggestions for research and operational work to help vulnerable people on the move in the face of climate and other risks.

Institute for Environmental & Human Security   Koko Warner
United Nations University
Bonn, Germany

References

Adger, W. N., Pulhin, J. M., Barnett, J., Dabelko, G. D., Hovelsrud, G. K., Levy, M., Oswald Spring, Ú., & Vogel, C. H. (2014). Human security. In C. B. Field, V. R. Barros, D. J. Dokken, K. J. Mach, M. D. Mastrandrea, T. E. Bilir, M. Chatterjee, K. L. Ebi, Y. O. Estrada, R. C. Genova, B. Girma, E. S. Kissel, A. N. Levy, S. MacCracken, P. R. Mastrandrea, & L. L. White (Eds.), *Climate change 2014: Impacts, adaptation, and vulnerability. Part A: Global and sectoral aspects. Contribution of Working Group II to the fifth assessment report of the Intergovernmental Panel on Climate Change* (pp. 755–791). Cambridge/New York: Cambridge University Press.

Afifi, T., Milan, A., Etzold, B., Schraven, B., Rademacher-Schulz, C., Sakdapolrak, P., Reif, A., van der Geest, K., & Warner, K. (2015). Human mobility in response to rainfall variability: Opportunities for migration as a successful adaptation strategy in eight case studies. *Migration and Development*. doi:10.1080/21632324.2015.1022974.

Feng, S., Krueger, A. B., & Oppenheimer, M. (2010). Linkages among climate change, crop yields and Mexico–US cross-border migration. *Proceedings of the National Academy of Sciences 107*(32), 14257–14262. Changes by 2030. *Global Environmental Change, 20*(4), 577–585.

Intergovernmental Panel on Climate Change (IPCC). (2012). In C. B. Field, V. Barros, T. F. Stocker, D. Qin, D. J. Dokken, K. L. Ebi, M. D. Mastrandrea, K. J. Mach, G.-K. Plattner, S. K. Allen, M. Tignor, & P. M. Midgley (Eds.), *Managing the risks of extreme events and disasters* to advance climate change adaptation. *A special report of Working Groups I and II of the Intergovernmental Panel on Climate Change* (p. 582). Cambridge/New York: Cambridge University Press.

Intergovernmental Panel on Climate Change (IPCC). (2014). Intergovernmental Panel on Climate Change: Climate change 2014: Impacts, adaptation, and vulnerability, Part A: Global and sectoral aspects. In C. B. Field, V. R. Barros, D. J. Dokken, K. J. Mach, M. D. Mastrandrea, T. E. Bilir, M. Chatterjee, K. L. Ebi, Y. O. Estrada, R. C. Genova, B. Girma, E. S. Kissel, A. N. Levy, S. MacCracken, P. R. Mastrandrea, & L. L. White (Eds.), *Contribution of Working Group II to the fifth assessment report of the Intergovernmental Panel on Climate Change* (p. 1132). Cambridge/New York: Cambridge University Press.

Klein, R. J. T., Midgley, G. F., Preston, B. L., Alam, M., Berkhout, F. G. H., Dow, K., & Shaw, M. R. (2014). Adaptation opportunities, constraints, and limits. In C. B. Field, V. R. Barros, D. J. Dokken, K. J. Mach, M. D. Mastrandrea, T. E. Bilir, M. Chatterjee, K. L. Ebi, Y. O. Estrada, R. C. Genova, B. Girma, E. S. Kissel, A. N. Levy, S. MacCracken, P. R. Mastrandrea, & L. L. White (Eds.), *Climate change 2014: Impacts, adaptation, and vulnerability. Part A: Global and sectoral aspects. Contribution of Working Group II to the fifth assessment report of the Intergovernmental Panel on Climate Change* (pp. 899–943). Cambridge/New York: Cambridge University Press.

Oppenheimer, M., Campos, M., Warren, R., Birkmann, J., Luber, G., O'Neill, B. C., & Takahashi, K. (2014). Emergent risks and key vulnerabilities. In C. B. Field, V. R. Barros, D. J. Dokken, K. J. Mach, M. D. Mastrandrea, T. E. Bilir, M. Chatterjee, K. L. Ebi, Y. O. Estrada, R. C. Genova, B. Girma, E. S. Kissel, A. N. Levy, S. MacCracken, P. R. Mastrandrea, & L. L. White (Eds.), *Climate change 2014: Impacts, adaptation, and vulnerability. Part A: Global and sectoral aspects. Contribution of Working Group II to the fifth assessment report of the Intergovernmental Panel on Climate Change* (pp. 1039–1099). Cambridge/New York: Cambridge University Press.

Röckstrom, J., Steffen, W., Noone, K., Persson, A., Chapin, F. S., III, Lambin, E., Lenton, T. M., Scheffer, M., Folke, C., Schellnhumber, H., et al. (2009). A safe operating space for humanity. *Nature, 461*, 472–475

Warner, K., & Afifi, T. (2014). Where the rain falls: Evidence from 8 countries on how vulnerable households use migration to manage the risk of rainfall variability and food insecurity. *Climate and Development, 6*(1), 1–17. doi:10.1080/17565529.2013.835707.

Part I
Mountain Areas

Chapter 1
An Index Based Assessment of Vulnerability to Floods in the Upper Indus Sub-Basin: What Role for Remittances?

Soumyadeep Banerjee, Muhammad Zubair Anwar, Giovanna Gioli, Suman Bisht, Saleem Abid, Nusrat Habib, Sanjay Sharma, Sabarnee Tuladhar, and Azeem Khan

1.1 Introduction

Mountain households tend to pursue a multi-income livelihood system, which combines farm and non-farm options. The non-farm strategies include wage employment, trade, and labor migration to varying degrees (Kreutzmann 1993). Labor migration can benefit recipient households through financial and social remittances.[1] The rationale behind remittances as a 'risk mitigation strategy' comes from growing evidence that they tend to be a counter-cyclical shock absorber in times of crisis (Agarwal and Horowitz 2002; Osili 2007).[2] In mountain contexts of the global South, lack of formal employment opportunities, precarious land rights, subsistence agriculture, along with the lack of access to financial instruments and social protection, severely limit the ability of people to cope with crisis and insure themselves against risks (e.g. economic, environmental, social, and political). The incomes from in-situ livelihood sources and labor migration are unlikely to be disrupted by

---

[1] Migrants facilitate a circulation of ideas, practices, and identities between destination and origin communities. These are referred as social remittances (Levitt 2001).

[2] Some studies also show that remittances can be pro-cyclical, because migrants' decision to remit is also driven by factors such as investment in physical and human capital (see e.g. Cooray and Mallick 2013).

S. Banerjee (✉)
International Centre for Integrated Mountain Development (ICIMOD), Kathmandu, Nepal

University of Sussex, Brighton, UK
e-mail: soumyadeep.banerjee@icimod.org

M.Z. Anwar • S. Abid • N. Habib • A. Khan
National Agricultural Research Centre (NARC), Islamabad, Pakistan

G. Gioli • S. Bisht • S. Sharma • S. Tuladhar
International Centre for Integrated Mountain Development (ICIMOD), Kathmandu, Nepal

© Springer International Publishing Switzerland 2016
A. Milan et al. (eds.), *Migration, Risk Management and Climate Change: Evidence and Policy Responses*, Global Migration Issues 6,
DOI 10.1007/978-3-319-42922-9_1

environmental hazards at the same time (Osili 2004). Financial remittances (hereafter remittances) have been described as a form of a household sponsored insurance system (Yang and Choi 2007; Paulson 2003), and a substitute for social security (Schrieder and Knerr 2000). Labor migration has shown to also enhance access to information and expand social networks (ADB 2012).

Over the past three decades, the framing of the nexus between migration and environmental change has shifted from that of securitization (Suhrke 1994; Myers 2002) with a narrow focus on environmental pull-factors and forced migration, to developmentalisation (Felli 2013; Bettini and Gioli 2015). Developmentalisation fosters the idea that *labor* migration could represent a legitimate and positive adaptation strategy to (global) environmental change (Foresight 2011; Warner et al. 2012). The 'migration as adaptation' thesis could be considered as a subset of the 'migration and development' discourse,[3] and is conceptually grounded on the merging of the New Economics of Labour Migration (Stark and Levhari 1982; Stark and Bloom 1985) and Sustainable Livelihoods Approach (Scoones 1998). This provides a framework for understanding mobility as a household strategy for managing various types of risk. Literature shows that remittances contribute to the welfare of vulnerable households by easing the fulfillment of basic needs such as purchasing food, housing, equipment, and paying for education and healthcare (Lindley 2009; Deshingkar 2006). Despite the fact that remittances seldom benefit the poorest households (Mazzucato et al. 2008) and may increase existing inequality at the micro level (Le De et al. 2013), econometric studies highlight that remittances may reduce the level and severity of poverty (World Bank 2012). Furthermore, the literature increasingly points at the crucial role of remittances during environmental disasters (for a review see Le De et al. 2013). For example, the IOM (2014:8) indicated that "in the context of natural or manmade catastrophes and crises, remittances and migration can support the resilience of populations both staying and going". Mohapatra et al. (2009) demonstrate that remittance recipient households in Ethiopia could use cash reserves to confront shocks to food security due to drought. Non-recipient households had to sell their livestock. Remittances also proved crucial for recovery in the aftermath of the 2004 Asian tsunami (Laczko and Collett 2005).

There is limited empirical evidence on relationships between environmental stressors, adaptation, and human mobility in Pakistan. A study by Mueller et al. (2014) that spanned over 21 years (1991-2012), linked individual-level information from the survey to satellite-derived measures of climate variability (Mueller et al. 2014). The findings indicate that in rural Pakistan heat stress has consistently increased long-term migration of men, driven by a negative effect on farm and non-farm income; whereas floods have a modest to insignificant impact on long-term migration. A study by Gioli et al. (2014) in the bordering region of Gilgit-Baltistan (i.e. Hunza and Yasin valleys), studied the role of migration and remittances in the

---

[3] The evolution of migration and development discourse has been discussed by de Haas (2012), and Gamlen (2014). For a discussion about similarities between the 'migration as adaptation' and 'migration and development' discourses, refer to Banerjee et al. (2012) and Bettini and Gioli (2015).

aftermath of two environmental shocks, the 2010 floods, and the landslide originating at the Attabad Lake in upper Hunza, considered as proxies for future climate impacts. This study found a high incidence of male circular labor migration (undertaken by 76 % of the surveyed households), occurring predominantly at the provincial and national level. The circular labor migration peaked in 2010 – the year in which the two environmental shocks had occurred – with 34 % of all the migrants' first migration occurring during 2010–2012 over a period spanning from 1985 to 2012. Many of these households had resorted to mobility as a coping mechanism in the aftermath of a shock, rather than as a proactive livelihood diversification strategy. Among those who lost their land (less than 1,500 m^2) and those unable to move (due to the lack of financial resources, employable skills, and human capital; family obligations; and illnesses) were significantly poorer (60 % less income) than the rest of the subsample. This highlights the need for aid agencies and governments to enhance outreach among the most vulnerable segments of society, i.e. those who are unable to 'self-insure' their lives through remittances (Gioli et al. 2014). Banerjee et al. (2011) focused on the relationship between water hazards and migration pattern in rural areas and the effect of remittances on the adaptive capacity of recipient households in four countries (Nepal, India, Pakistan and China) across the Hindu Kush Himalayan region (Pakistan: Chitral District, Khyber-Pakhtunkhwa Province). The likelihood that household members would migrate for work is higher among rural communities exposed to rapid onset water hazards (e.g. riverine or flash floods) than those exposed to slow onset water hazards (e.g. drought). The likelihood of labour migration is higher among households located in rural communities affected by very severe drought compared to households in less severely affected rural communities.

The increasing literature on remittances in times of crises indicates that remitted assets have a significant role in the aftermath of natural disasters (Bettin et al. 2014), and influence the vulnerability of remittance recipient households. The IPCC (2014:28) defines vulnerability as "the propensity or predisposition to be adversely affected. Vulnerability encompasses a variety of concepts including sensitivity or susceptibility to harm and lack of capacity to cope and adapt. The extent to which remittances can contribute to reduction of vulnerability requires further exploration. Remittances may be a significant source of disposable cash during a crisis; however, their role in reducing the vulnerability of a household by building medium term and long term assets is little understood. Past research (de Haan 1999; de Haas 2012) suggests that the development outcome of migration is context dependent. Therefore, it is likely that the effect of remittances on vulnerability will be context dependent. This context is critical to vulnerability as well. For example, livelihoods in mountain regions are often characterized by a lack of insurance and formal measures of social protection. As elsewhere, livelihoods vulnerability is embedded in everyday power relations influenced by class (Mustafa 2005; Pelling 1998), gender (Sultana 2010), and ethnicity (Bolin 2007) among other factors (*social vulnerability*). There is a high prevalence of family farming and livestock rearing and widespread dependence on natural resources, all of which are highly sensitive to environmental and climatic changes (*biophysical vulnerability*). Mountain populations are often marginalized in terms of political participation and inclusion in political and institution building processes (*political vulnerability*, see Wisener and Luce 1993).

This chapter analyses the vulnerability of remittance recipient and non-recipient households in the flood-affected rural communities of Upper Indus Sub-Basin (UISB). A vulnerability assessment has been conducted to characterize the adaptive capacity, exposure, and sensitivity of remittance recipient and non-recipient households. This chapter will explore whether the remittances have a role in reducing the vulnerability of households to floods, by attempting to answer the following question: How is vulnerability of remittance recipient households to floods different from that of non-recipient households? We adopt an index based approach to explore the aforementioned question. It provides a metric for quantitative analysis of a household's vulnerability to a specific environmental stressor. This chapter is organized as follows. The next section provides an overview of the research methodology, research method, and study area. Then, we present empirical evidence in order to characterize vulnerability in remittance recipient and non-recipient households. Finally, the policy implications of these findings are discussed.

1.2 Indices of Vulnerability at the Household Level

1.2.1 Research Methodology

A diversity of methodologies have been used to assess vulnerability. These include simulation based models (e.g. Brenkert and Malone 2005), indicator based approaches (e.g. Vincent 2007) and participatory exercises (e.g. Gupta et al. 2010). These methodologies have been applied to different systems or spatial scales of analysis: district (e.g. Hahn et al. 2009), community (e.g. Pelling and High 2005), sector (e.g. Eakin et al. 2011), and particular ecosystem (e.g. Shah et al. 2013). Secondary data (e.g. Brooks et al. 2005), primary data from household surveys (e.g. Hahn et al. 2009), and participatory exercises (e.g. Gupta et al. 2010) have been used to explore the aforementioned methodologies. In this study, vulnerability is conceptualized to be a function of three major components, namely adaptive capacity, exposure, and sensitivity. These major components are composed of sub-dimensions, which are comprised of attributes that can be measured through specific indicators. Adaptive capacity of a household is comprised of sub-dimensions such as financial asset, physical asset, natural asset, social asset, and human asset. Financial asset is represented by access to formal financial institutions and access to insurance. Natural asset is comprised of farm size, number of livestock, and changes in agricultural practice due to flood. Access to flood assistance, access to borrowing during floods, and participation in collective action for flood relief, recovery, and preparedness are attributes of social asset. Human asset is comprised of access to information, access to local alternative livelihood opportunities, and access to alternative livelihood opportunities in nearby localities. Physical asset has three attributes, namely structural changes in houses to address flood impacts, access to storage options during floods, and farm mechanization.

Exposure of a household to floods is comprised of three sub-dimensions: Average financial damage to a household due to floods between 1984 and 2013, number of floods experienced by a household between 1984 and 2013, and average time required by a household to recover from the damages it had experienced due to flood. Sensitivity of a household to floods is comprised of well-being, water, food, and environmental dependence. The well-being sub-dimension is represented by reduction in spending on education and clothes due to flood, selling or mortgaging of household assets, and sending children to work outside the household as a result of floods. The water sub-dimension is comprised of the average time taken by a household member to collect drinking water for a normal day, lack of drinking water storage for emergency consumption during floods, and lack of arrangements to treat drinking water for consumption during floods. The food sub-dimension includes reliance on less preferred food items during floods, restricted food consumption by adults due to floods, not spending savings to procure food during floods, begging for food due to floods, and not collecting wild food due to floods. The environmental dependence sub-dimension is comprised of dependence on subsistence farming, crop diversification, dependence of household income on the primary sector, reduction in agricultural assets due to floods, dependence on environmental resources for the primary source of cooking fuel, and households with less resistant construction material for external walls.

This study has adopted the equal weighted design to construct the vulnerability index. Hahn et al. (2009) had assigned equal weight to all the indicators based on the assumption that all are of equal importance. Each major component, sub-dimension, and attribute contributes equally to the overall index. Once each of the attributes are standardized, they are averaged using Eq. 1.1, to calculate the value of each sub-dimension:

$$S_s = \frac{\sum_{i=1}^{n} index_{a_s}}{n} \qquad (1.1)$$

where S_s is one of the sub-dimensions of sensitivity, exposure, or adaptive capacity for a household in particular study area s. For example, sensitivity is comprised of sub-dimensions such as well-being, water, food, and environmental dependence, and n is the number of attributes in each sub-dimension. After value of each sub-dimension is calculated, they are averaged using Eq. 1.2 to obtain the major components, i.e. Sensitivity index (*SI*), Exposure index (*EI*), and Adaptive Capacity Index (*AI*):

$$M_s = \frac{\sum_{i=1}^{n} w_i S_{si}}{\sum_{i=1}^{n} w_i} \qquad (1.2)$$

where M_s is a major component of vulnerability (i.e. sensitivity, exposure, or adaptive capacity) for a household in particular study area s, weight w_i is determined by the number of sub-dimensions that contribute to a major component, and S_{si} is average value of sub-dimensions comprising each of the major components. The three major components were combined using the following equation:

$$VI_s = (EI-AI)-SI \qquad (1.3)$$

Where VI_s is the vulnerability index for a household in a particular study area s, EI, AI, and SI are the exposure index, the adaptive capacity index, and the sensitivity index for the same household. The VI ranges from −1 to +1.

According to the New Economics of Labour Migration (NELM), migration is a risk sharing strategy of the household to diversify resources in order to minimize income risks (Stark and Levhari 1982). By broadening the space through migration of one or more household members in search of employment, a household attempts to overcome constraints to its development (Stark and Bloom 1985; Stark and Lucas 1988), weakly developed credit and insurance markets (Taylor 1999), and invest in productive activities and improve their livelihoods (de Haas 2007). The costs and returns of migration are shared by the migrant and sending household (Stark and Bloom 1985; Stark and Lucas 1988). Remittances maintain a functional linkage between the migrant worker in the destination area and the migrant-sending household in the origin community. In this study, a household was considered to be a remittance recipient household if it had received remittances from any household member who had lived and worked in another village or town in the same country or another continuously for 2 months or more at any time during the last 30 years. A household not conforming to this definition was considered to be a non-recipient household. Further elaboration of the research methodology could be found in Banerjee et al. (forthcoming) and Banerjee (forthcoming).

1.2.2 Research Method

A survey was conducted in Hunza, Ghizer, Gilgit, and Chitral districts from October 2014 to March 2015. These districts are considered as one aggregated areal unit, and are representative of the Upper Indus Sub-basin (UISB). The survey gathered primary data on socio-demographic characteristics (age, gender, ethnicity, educational level of the household head), household assets (financial, human, physical, natural, social), prior occurrence and economic damages due to floods between 1984 and 2013, flood response strategies, food and non-food expenditure, livelihood practices (access to land, types of crops grown, livestock rearing, labour migration), and income sources. A list of all flood affected villages was prepared for the study area.[4] The selection of households involved a two stage process. First, villages are selected using the Probability Proportional to Size (PPS). Second, an equal number of households is selected using systematic sampling within each selected

---

[4] If a village had experienced a riverine flood or flash flood at least once since 1984, it was considered as flood affected.

1 An Index Based Assessment of Vulnerability to Floods in the Upper Indus... 9

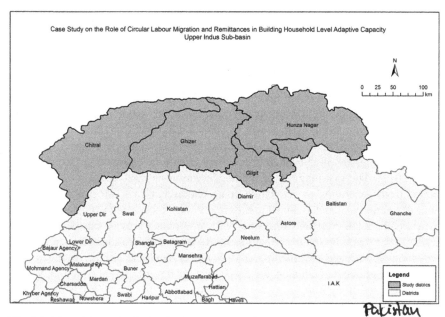

Fig. 1.1 Map of the study districts in the Upper Indus Sub-basin (Source: Migration Case Study, Himalica programme, ICIMOD)

village. A sample size of 360 households is estimated, 180 each for remittance recipient and non-recipient households.[5] The primary sampling unit was 20 households (10 each for remittance recipient and non-recipient households) in each village and therefore, 18 villages were selected. At the end of the survey, a sample size of 358 was achieved; 179 remittance recipient households and 179 non-recipient households.

1.2.3 Description of the Study Area in the Upper Indus Sub-Basin (UISB)

The study area lies in the UISB, where the observed climate trends are anomalous. As opposed to the Eastern Himalayas, the UISB is experiencing since decades cooling trends in the summer season, and increasing or stable precipitations throughout the year (Fowler and Archer 2006; Bocchiola and Diolaiuti 2012), accompanied by mass gains in the glaciers of the Karakoram (Bolch et al. 2012). This case study covers four districts in the UISB: Hunza, Ghizer, Gilgit, and Chitral (see Fig. 1.1). Hunza, Gilgit, and Ghizer are located in the region of Gilgit-Baltistan (Eastern

---

[5] If at any time during the past 30 years a household had received financial remittances from any household member who had lived and worked in another village or town in the same country or another continuously for 2 months or more, it was referred as a remittance recipient household.

Karakoram). Chitral is located in the Khyber-Pakhtunkhwa province (Hindu-Kush). Despite belonging to different politico-administrative units, these areas have similar physical and socio-economic characteristics, and since the 1980s have followed comparable patterns of development, largely resulting from the implementation of the Aga Khan Rural Support Program (AKRSP)[6] model. These valleys are characterized by an extreme environment and an arid climate. Agricultural production is made possible by the high incidence of solar radiation, and a complex indigenous irrigation system relying on melt-water channeled directly from the glaciers to the flat areas at the bottom of the valleys. These irrigation oases (less than 1 % of the Karakoram region) offer limited space for agricultural production (Kreutzmann 1993, 2011). The vast majority of the population owns small pieces of land transmitted from generation to generation along patriarchal lines. Most of the grazing areas are communal and assigned to different villages according to customary laws. Wheat is the main crop, and since the 1970s it has been heavily subsidized (in the form of tax exemption) by the Government of Pakistan. There has been a significant reduction in per capita availability of agricultural land and grazing pastures as a result of the growing population and environmental hazards. Local communities are increasingly shifting from an agropastoral economy to a combined subsistence-labor system (Ehlers and Kreutzmann 2000). Within the latter system, the households pursue risk prone mountain agriculture with external income-generating opportunities, such as labor migration, wage labor, and trade. The non-farm income from external sources has been facilitated by pivotal infrastructure development, such as the Karakoram Highway. Rising levels of education have also contributed to increasing the share of people employed in governmental jobs and in the tertiary sector (Malik and Piracha 2006). Yet, most households cultivate land and rear livestock on a small scale.

1.3 Results

1.3.1 Livelihoods Portfolio

This section provides an overview of the livelihoods portfolio of remittance recipient and non-recipient households. The majority of surveyed households (92 %) have access to farm land. On average, a household has 0.84 ha (remittance recipient: 0.94 ha, non-recipient: 0.76 ha). Remittance recipient households own almost all of the farm land to which they had access (97 %). Non-recipient households own approximately two-thirds of their farm land (67 %), and have mainly leased the rest. These households grow wheat, maize, and summer vegetables. Some of the households also grow apple, apricot, and walnut. Average income from crop sales for

---

[6] Since the 1980s, various NGOs and in particular the Aga Khan Development Network (AKDN) and its Aga Khan Rural Support Program (AKRSP) have introduced cash crops such as potatoes and orchards (mostly almonds, apricots, grapes, and cherries).

remittance recipient and non-recipient households during the year preceding the survey is estimated to be USD 981 and USD 1,847 respectively. The economic status of the household is represented by the average monthly per capita expenditure (MPCE) of the households, which comprises food and non-food expenditure. Remittance recipient and non-recipient households in the bottom expenditure category ('low income households') sell staple crops. The average income from the sale of staple crops during the 12 months preceding the survey among remittance recipient and non-recipient households in the bottom expenditure category was USD 2,298 and USD 1,412 respectively. In the top expenditure category ('high income households'), sale of staple crops is largely limited to non-recipient households. The average income from sale of staple crops among non-recipient households in the top expenditure category (USD 3,189.06) is far higher than that of remittance recipient households in the same category (USD 143.03). This indicates that the sale of staple crops is a means of acquiring cash to address basic needs and urgent necessities among non-recipient and 'low income' remittance-recipient households. Most of the households have access to livestock (96 %), but less than 2 % of households have reported the sale of livestock and livestock products as a major source of household income.

Contribution of salaried employment from local non-farm sources, business or trade, and daily wages from local non-farm sources vary greatly among recipient and non-recipient households. Non-recipient households have better access to in-situ non-farm livelihood opportunities such as salaried employment, daily wage, and small business. Salaried employment from non-farm sources in the locality is a major source of income for one-third of non-recipient households and one-fifth of remittance recipient households. Another one-fifth of non-recipient households have reported daily wages from non-farm sources in the locality as a major source of household income; compared to that of less than one-twentieth in remittance recipient households. Around one fourth of non-recipient households identified business or trade as their major source of income, while only one-tenth of remittance recipient households have reported business or trade as the major source of household income. On the other hand, two-thirds of remittance recipient households and a quarter of non-recipient households have a household member who commutes to work. These commuters are men of working age, and employed in the defense, public administration and education sectors. Overall, non-recipient households have better access to local non-farm income opportunities than remittance recipient households.

Remittance recipient households have substituted the local non-farm income sources with remittances, which is the major income source for one-third of remittance recipient households (see Fig. 1.2). Migration from the study area is predominantly internal in nature. Among 364 migration episodes between 1984 and 2013, over three quarters are associated with an urban destination (84 %) in Pakistan. Popular migration destinations are located in Gilgit-Baltistan, Punjab, Khyber-Pakhtunkhwa, and Sindh. A large number of these migrant workers are employed in the formal sector. For example, over half of the surveyed migrant workers are covered by social security benefits (e.g. pension, provident fund, or insurance) or receive

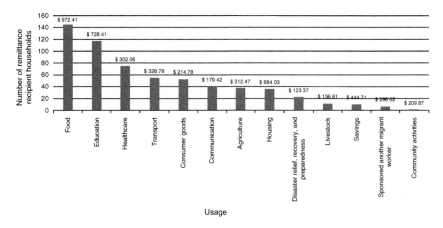

Fig. 1.2 Use of the financial remittances during the 12 months preceding the survey, Upper Indus Sub-basin, 2014–2015 (Source: Migration Case Study, Himalica programme, ICIMOD)

paid leave as part of their latest job in destination. This could be explained by the fact that over half of the surveyed migrant workers are employed in the defense sector. During the 12 months preceding the survey, the mean volume of domestic and foreign remittances was estimated to be USD 1,590 and USD 430 respectively. Further analysis of MCPE categories indicates that mean volume of domestic remittances progressively increased from households in bottom category through middle to top category. Around nine-tenths of the households receive remittances in the form of cash, which indicates that hand-carrying by migrant worker or an acquaintance is the preferred mode of remittance transfer. In addition, one-third of households had also received remittances in form of consumer goods, and one-tenth of households had received remittances through bank cheques. During the 12 months preceding the survey, remittances were mainly used to meet food and educational expenditure (see Fig. 1.2). Other common usages of remittances include healthcare, transport, and consumer goods. There is little investment of remittances in agriculture, housing, disaster risk reduction, savings, livestock rearing, and community activities.

1.3.2 Floods and Flood Responses

Floods are a major environmental stressor in the study area. The survey data indicates that between 1994 and 2013, half of the households had experienced two flood events, one-fourth of households had experienced four flood events, one-tenth of households had experienced three flood events, and remaining households had experienced one flood event. During the same period, on average, remittance recipient and non-recipient households had incurred flood damages of USD 4,733 and USD 3,584, respectively. However, village level flood preparedness remains low. For example, most of the study villages do not have a pre-designated flood shelter

Table 1.1 Top five household level flood responses during, in immediate aftermath, and between floods, Upper Indus Sub-basin, 2014–2015

Top five actions taken due to floods

	During flood	Immediate aftermath of flood	Between two flood events
Action 1	Moved family to a safer location (29.7 %)	Cleared or reconstructed the water channel (26.6 %)	Reconstructed the water channel (12.9 %)
Action 2	Moved cattle to a safer location (27.8 %)	Prepared to farm (22.7 %)	Re-established the orchard (7.5 %)
Action 3	Prayed to God for safety (22.9 %)	Re-established orchard (16.5 %)	Prepared to farm (9.3 %)
Action 4	Stored valuables in a safe place (5.4 %)	Cleaned or repaired the house (14.5 %)	Raised plinth of the house (9.3 %)
Action 5	Stored grains in a safe place (5.5 %)	Rebuilt the house (14.1 %)	Built a barrier to arrest debris flowing in water (5.4 %)

Source: Migration Case Study, Himalica programme, ICIMOD

for either people or livestock or a village level flood contingency plan, nor do they discuss flood preparedness during village meetings. The household level flood responses are mostly short-term in nature. The household level flood responses could be distinguished as responses during the flood inundation, in the immediate aftermath of a flood (when flood water has receded), and between two distinct flood events (see Table 1.1). During the flood inundation, common household responses focus on evacuation and relief measures such as moving the family and cattle to a safer location. In the immediate aftermath of a flood, households engage in recovery, which includes cleaning or reconstructing water channels, preparing farms for cultivation, re-establishing orchards, repairing houses or cattle-sheds, and arranging safe drinking water. The major household level flood response strategies between two flood events are similar to the strategies adopted in its immediate aftermath. These strategies are short-term in nature, focus on immediate necessities, and help the households to cope with flood impacts. These do not address the underlying factors that make households vulnerable to flood impacts. On average, households adopt four flood response strategies.

1.3.3 Vulnerability Index

The vulnerability of remittance recipient household (RRHH) is marginally lower than non-recipient households (NRHH) (vulnerability index: NRHH: â~0.22, RRHH: â~0.20). There are some differences between these two groups of household at the levels of sub-dimensions or attributes (see Table 1.2). There is marginal difference between sensitivity of remittance recipient and non-recipient households

Table 1.2 Sub-dimensions and attributes of sensitivity by remittance recipient status of the household, Upper Indus Sub-basin[a]

Sub-dimension	Non-recipient households	Recipient households	Attribute	Non-recipient households	Recipient households
Well being	0.03	0.02	Reduced educational spending due to floods	0.05	0.02
			Reduced spending on clothes due to floods	0.06	0.03
			Sent children to work outside the household due to floods	0.01	0.02
			Sold or mortgaged household assets due to floods	0.01	0.01
Water	0.64	0.63	Average time taken by a member of your household to collect drinking water for a normal day	0.05	0.06
			Did not store drinking water for consumption during inundation	0.97	0.95
			Did not arrange safe drinking water or treat drinking water for consumption during inundation	0.91	0.89

(continued)

Table 1.2 (continued)

Sub-dimension	Non-recipient households	Recipient households	Attribute	Non-recipient households	Recipient households
Food	0.44	0.42	Relied on less preferred food items due to flood	0.1	0.02
			Restricted food consumption of adults due to flood	0.23	0.16
			Did not collect wild food during flood	0.99	0.99
			Did not spend savings to procure food during flood	0.87	0.92
			Begged for food due to flood	0.02	0.01
Environmental dependence	0.52	0.49	Dependent on subsistence farming	0.12	0.12
			Crop diversification	0.79	0.65
			Dependent on primary sector for household income	0.09	0.05
			Reduction in agricultural assets due to floods	0.05	0.05
			Household with less resistant construction material for the external wall	0.86	0.84
			Dependent on environmental resources for the primary source of cooking fuel	0.95	0.97

Source: Migration Case Study, Himalica programme, ICIMOD
[a]The sub-dimensions and attributes have been standardized

Table 1.3 Sub-dimensions of exposure by remittance recipient status of the household, Upper Indus Sub-basin[a]

Sub-dimension	Non-recipient households	Recipient households
Flood frequency between 1984 and 2013	0.30	0.32
Damage due to floods between 1984 and 2013	0.01	0.03
Recovery from flood damage between 1984 and 2013	0.01	0.01

Source: Migration Case Study, Himalica programme, ICIMOD
[a]The sub-dimensions and attributes have been standardized

(sensitivity index: NRHH: 0.41, RRHH: 0.39). Though a reduction in educational spending due to floods is low among the households in the study area, this reduction is marginally higher among non-recipient households than remittance recipient households. During floods, non-recipient households rely more on less preferred food items and reduce number of meals among adult household members than remittance recipient households. Environmental dependence of non-recipient households is higher than remittance recipient households. Among the households engaged in farming, crop diversification is higher among remittance-recipient households than among non-recipient households (Table 1.3).

The adaptive capacity of remittance recipient households is marginally higher (0.62) than non-recipient households (0.64) (see Table 1.4).[7] Remittance recipient households have better access to financial assets than non-recipient households. Particularly, remittance recipient households have a better access to formal financial institutions (e.g. savings bank account). Little less than nine-tenths of remittance recipient households have a bank account; while two-thirds of the non-recipient households have one. However, few remittance recipient and non-recipient households undertook targeted savings as a strategy to manage environmental risks, particularly those posed by floods. The insurance penetration remains low in the UISB. Life insurance is the common type of insurance, which is available in one-third of remittance recipient and non-recipient households. Less than one-tenth of these households have a health insurance. None of the surveyed households have either crop or livestock insurance, which could be a useful option to manage environmental risks. More remittance recipient households have access to alternative livelihood opportunities in a nearby locality ('commute') than non-recipient households. Among the households engaged in farming, more non-recipient households have made changes in agricultural practices in response to floods than remittance recipient households. Remittance recipient households have lower access to borrowing during the floods than non-recipient households.

---

[7] Higher the value of adaptive capacity index, lower is the adaptive capacity.

Table 1.4 Sub-dimensions and attributes of adaptive capacity by remittance recipient status of the household, Upper Indus Sub-basin[a]

Sub-dimension	Non-recipient households	Recipient households	Attribute	Non-recipient households	Recipient households
Financial asset	0.51	0.40	Did not have access to formal financial institution	0.37	0.15
			Did not have access to insurance	0.66	0.65
Natural asset	0.60	0.60	Farm size diversification	0.73	0.69
			Livestock diversification	0.18	0.19
			Did not make changes in agricultural practices due to floods	0.87	0.93
Social asset	0.67	0.69	Did not receive flood assistance	0.54	0.50
			Did not have access to borrowing during floods	0.79	0.86
			Did not participate in collective action on flood relief, recovery, or preparedness	0.69	0.70
Human asset	0.67	0.64	Communication device diversification index	0.38	0.40
			Did not have access to local alternative livelihoods opportunity	0.89	0.91
			Did not have access to alternative livelihoods opportunity in nearby locality	0.75	0.62
Physical asset	0.71	0.69	Did not make structural changes in the house due to flood	0.58	0.58
			Did not have access to farm mechanization	0.92	0.93
			Did not have access to storage options during flood	0.64	0.56

Source: Migration Case Study, Himalica programme, ICIMOD
[a]The sub-dimensions and attributes have been standardized

1.4 Conclusion: Mainstreaming Human Mobility in Climate Adaptation Policy

The nexus between migration and adaptation had officially entered the global adaptation agenda in 2011, with the inclusion of migration in the Â§14f of the CancÃºn Framework for Adaptation. Since then, several steps have been taken towards recognizing the potential role of mobility in fostering resilience. Migration features in the Warsaw International Mechanism for the definition of 'Loss and Damage' (James et al. 2014). The Intergovernmental Panel on Climate Change (IPCC 2014) has devoted an increased attention to the matter, notably in Chapter 12 of its latest assessment report. The Paris COP21 agreement includes migration and human mobility in the Preamble, and displacement in the section on Loss and Damage (paragraph 50). The Preamble of the agreement states that the "Parties should, when taking action to address climate change, respect, promote and consider their respective obligations on human rights, the right to health, the rights of indigenous peoples local communities, *migrants*, children, persons with disabilities and people in vulnerable situations and the right to development, as well as gender equality, empowerment of women and intergenerational equity (UNFCCC 2015:1)".[8] Paragraph 50 requests the Executive Committee of the Warsaw International Mechanism to establish a task force that will develop recommendations for integrated approaches to avert, minimize, and address displacement related to the adverse impacts of climate change (UNFCCC 2015). However, when it comes to implementation at the national and sub-national levels, migration continues to be framed negatively, and is predominately understood as an undesired outcome of (global) environmental change, i.e. as a last resort option signaling a failure to adapt.

Despite the importance of mobility and its pivotal role in the Pakistan's basket of livelihoods and (macro) economic dynamics, migration has been portrayed as a problem in the national debate on adaptation. Pakistan has approved a National Climate Change Policy (NCCP) in 2012 (GoP 2012). The NCCP does not mention the nexus between mobility and climate change adaptation (CCA). It has a narrow understanding of migration. For instance, rural-to-urban migration is characterized as a threat to urban development. The NCCP aims to develop "measures to curb rural-to-urban migration, develop infrastructure and support facilities in smaller agro-based towns and periphery urban areas" (GoP 2012: 26). The complex urbanization dynamics – which are integral part of wider development processes – are described as something that must be stopped rather than governed through the right set of policies. Nonetheless, the document acknowledges the need to "combine efforts towards poverty alleviation along with management of climate change impacts and environmental degradation effects (...), and to integrate the poverty-climate change nexus into economic policies and plans" (GOP 2012: 25). This means to foster sustainable rural and urban development and creating job opportunities (livelihood security). With support from policy and institutional mechanisms, financial and social remittances can play an important role in CCA. Empirical evi-

---

[8] Italicized by the authors.

dence would help to enhance understanding among the national stakeholders about the relationship between migration and CCA, particularly at the household level.

Floods are a major environmental stressor in the study area. Yet, village level flood preparedness remains low, and household level flood preparedness is comprised of short-term strategies. Remittances are crucial to meet the basic needs of recipient households. However, there is little investment of remittances in agriculture, housing, savings, or disaster risk reduction. It is evident that flood preparedness is not a priority of either remittance recipient or non-recipient households.

Based on the evidence of this study, the following section discusses policy options to support remittance recipient households to better manage risks from environmental stressor. The local government institutions, development partners, and community-based organizations could raise awareness about long term benefits of disaster risk reduction, particularly among women, children, and youth. The autonomous initiatives at the household level could be nurtured by government and non-government institutions. The awareness raising campaign could be supplemented by technical inputs that the households may require to build capacity or deploy strategies intended to manage environmental risks. The role of labor migration and remittance should be mainstreamed in the NCCP and associated policy instruments. Potential ways to support the use of remittances to reduce disaster risk among remittance recipient households (e.g. safe drinking water, housing, food security, health care, veterinary care, or livelihoods diversification) could be explored by the local government institutions as part of development and adaptation planning. There is a high degree of uncertainty about the future direction and magnitude of change in the UISB. Hence, adaptation planning should focus on 'no-regret' strategies, which aim to increase robustness, enhance flexibility, ensure efficiency, and guarantee equity. The national discourse on environmental hazards, adaptation, and migration is at a nascent stage. Even within this rudimentary discourse, the main focus is on the influence of climate change on migration decision-making. As indicated in this case study, internal migrant workers are an important constituent of the migrant population in the UISB. Yet, the internal migrant workers are missing in the national discourse on environmental hazards, adaptation, and migration. A coordinated awareness-raising programme would be required to raise awareness of stakeholders such as government institutions, non-government organizations, community based organizations, and financial institutions, about the potential role of remittances within existing CCA and DDR policies.

The findings from the vulnerability assessment indicate that remittance recipient households are marginally less vulnerable than non-recipient households. The sensitivity of remittance recipient households is a little lower than non-recipient households and the adaptive capacity of the former is marginally higher. The dependency of non-recipient households on the environmental resources is higher than the remittance recipient households. Therefore, it could be inferred that the livelihoods portfolio of non-recipient households is at a higher risk from environmental stressors. For example, the sale of staple crops remains a major source of cash income for the low and high income non-recipient households. In an increasingly monetized rural economy, cash income is necessary to meet the basic needs of a household.

Availability of cash is even more critical for the sustenance of a household during a crisis. If production of staple crops is vulnerable to the environmental stressors (including floods), the welfare of the household is at risk. Local government institutions could raise awareness about the long term benefits livelihoods diversification among the low income non-recipient households, and facilitate access to non-farm income opportunities. The households in the study area have limited access to formal insurance mechanisms. A household that possesses either a crop or livestock insurance is a rarity. The government institutions and international donors could support the expansion of formal insurance coverage in the study area, lower the cost of insurance premiums, simplify the paperwork, and reduce the time required to process a policyholder's claim. The expansion of an insurance network should be complemented by an awareness raising campaign among the potential beneficiaries and a 'last-mile' network of insurance agents to connect the households in remote rural areas. Remittance recipient households have better access to formal financial institutions. However, like non-recipient households, they do not undertake targeted savings to address future emergencies, including environmental risks. Existing financial literacy programs in the UISB could be revisited to incorporate modules on environmental risks and options to manage these risks (e.g. insurance, targeted savings, livelihood diversification, storage, financial inclusion, and disaster preparedness). Furthermore, financial institutions and development partners could explore means to provide customized financial products that could stimulate savings among the local population, particularly low income households, women, and youth.

Acknowledgement The authors express their gratitude to the editors for providing the opportunity to prepare this chapter. Particular appreciation goes to Dr. Koko Warner (UNU-EHS), Dr. Benjamin Schraven, (DIE), Ms. Noemi Cascone (UNU-EHS) and Dr. Andrea Milan (UNU-EHS) for their support, encouragement and insightful feedback. The authors would like to thank Dr. Golam Rasul (ICIMOD) and Mr. Valdemar Holmgren (ICIMOD) for their support and encouragement. Special thanks to Dr. Bidhubhusan Mahapatra (ICIMOD) for guidance with the statistical analysis. Particular appreciation also goes to Dr. Abdul Wahid Jasra (ICIMOD), Ms. Kanwal Waqar (ICIMOD), Mr. Muhammad Ayub (CESLAM), and the SSRI team for the invaluable support during the fieldwork. The authors would like to thank Dr. Vishwas Chitale (ICIMOD) and Mr. Gauri Dangol (ICIMOD) for their inputs in the preparation of this manuscript. The authors will like to thank the anonymous reviewers for their helpful comments. This research was supported by the Rural Livelihoods and Climate Change Adaptation in the Himalayas (Himalica) Programme, which is implemented by the International Center for Integrated Mountain Development (ICIMOD), and is funded by the European Union. The opinions expressed in the chapter are those of the authors and do not necessarily reflect the views of their respective institutions.

References

Agarwal, R., & Horowitz, A. (2002). Are international remittances altruism or insurance? Evidence from Guyana using multiple-migrant households. *World Development, 30*(11), 2033–2044.

Asian Development Bank. (2012). *Addressing climate change and migration in Asia and the Pacific*. Mandaluyong City: Asian Development Bank.

Banerjee, S., Gerlitz, J. Y. S., & Hoermann, B. (2011). *Labour migration as a response to water hazards in the Hindu-Kush-Himalayas*. Kathmandu: International Center for Integrated Mountain Development (ICIMOD).

Banerjee, S., Black, R., & Kniveton, D. (2012). *Migration as an effective mode of adaptation to climate change*. This paper was commissioned by the Foresight Committee (Go Science, UK) as a contribution to European Commission policy reflection on climate change and migration.

Banerjee, S. (forthcoming). *Understanding the effects of labour migration on vulnerability to extreme events in Hindu Kush Himalayas: Case studies from Upper Assam and Baoshan County*. D.Phil. thesis submitted to the University of Sussex.

Banerjee, S., Kniveton, D., Black, R., Bisht, S., Mahapatra, B., Tuladhar, S., & Das, P. J. (forthcoming). Does financial remittance build household level adaptive capacity? A case study of the flood affected households of Upper Assam in India. *Vulnerability and resilience in explaining migration and development*. Washington, DC: KNOMAD.

Bettin, G., Presbitero, A., & Sapatafora, N. (2014). *Remittances and vulnerability in developing countries* (Working Paper 14/13). International Monetary Fund.

Bettini, G., & Gioli, G. (2015). Waltz with development: Insights on the developmentalization of climate-induced migration. *Migration and Development*, doi:http://dx.doi.org/10.1080/216323 24.2015.1096143

Bocchiola, D., & Diolaiuti, G. (2012). Recent (1980–2009) evidence of climate change in the Upper Karakoram, Pakistan. *Theoretical and Applied Climatology, 113*, 611–641. doi:10.1007/s00704-012-0803-y.

Bolch, T., Kulkarni, A., KÃ¤Ã¤b, A., Huggel, C., Paul, F., Cogley, J. G., Frey, H., Kargel, J. S., Fujita, K., Scheel, M., Bajracharya, S., & Stoffel, M. (2012). The state and fate of Himalayan glaciers. *Science, 336*, 310–314. doi:10.1126/science.1215828.

Bolin, B. (2007). Race, class ethnicity and disaster vulnerability. In H. Rodriguez, E. L. Quarantelli, & R. R. Dynes (Eds.), *Handbook of disaster research* (pp. 113–129). New York: Springer.

Brenkert, A. L., & Malone, E. L. (2005). Modeling vulnerability and resilience to climate change: A case study of India and Indian states. *Climatic Change, 72*(1–2), 57–102.

Brooks, N., Adger, W. N., & Kelly, P. M. (2005). The determinants of vulnerability and adaptive capacity at the national level and the implications for adaptation. *Global Environmental Change, 15*(2), 151–163.

Cooray, A., & Mallick, D. (2013). International business cycles and remittance flows. *The B.E. Journal of Macroeconomics, 13*(1), 1–33.

de Haan, A. (1999). Livelihoods and poverty: The role of migration a critical review of the migration literature. *The Journal of Development Studies, 36*(2), 1–47. doi:10.1080/00220389908422619.

de Haas, H. (2007). *Remittances, migration and social development: A conceptual review of the literature* (Social Policy and Development Programme Number 34). Geneva: United Nations Research Institute for Social Development.

de Haas, H. (2012). The migration and development pendulum: A critical view on research and policy. *International Migration, 50*(3), 8–25. doi:10.1111/j.1468-2435.2012.00755.x.

Deshingkar, P. (2006). *Remittances in crisis: Sri Lanka after the Tsunami* (Humanitarian Policy Group Background Paper). London: Overseas Development Institute.

Eakin, H., Bojórquez-Tapia, L. A., Diaz, R. M., Castellanos, E., & Haggar, J. (2011). Adaptive capacity and social environmental change: Theoretical and operational modeling of smallholder coffee systems response in Mesoamerican Pacific Rim. *Environmental Management, 47*(3), 352–367.

Ehlers, E., & Kreutzmann, H. (2000). High mountain pastoralism in Northern Pakistan. *Erdkundliches Wissen, 132*, 9–36.

Felli, R. (2013). Managing climate insecurity by ensuring continuous capital accumulation: 'Climate refugees' and 'climate migrants'. *New Political Economy, 18*, 337–363.

Foresight. (2011). *Final project report – Foresight: Migration and global environmental change*. London: Government Office for Science.

Fowler, H. J., & Archer, D. R. (2006). Conflicting signals of climatic change in the Upper Indus Basin. *Journal of Climate, 9*, 4276–4293.

Gamlen, A. (2014). The new migration-and-development pessimism. *Progress in Human Geography, 38*(4), 581–597. doi:10.1177/0309132513512544.

Gioli, G., Khan, T., Bisht, S., & Scheffran, J. (2014). Migration as an adaptation strategy and its gendered implications: A case study from the Upper Indus Basin. *Mountain Research and Development, 34*(3), 255–265.

Government of Pakistan. (2012). *National climate change policy.* Islamabad: Ministry of Climate Change.

Gupta, J., Termeer, C., Klostermann, J., Meijerink, S., van den Brink, M., Jong, P., et al. (2010). The adaptive capacity wheel: A method to assess the inherent characteristics of institutions to enable the adaptive capacity of society. *Environmental Science and Policy, 13*(6), 459–471.

Hahn, M. B., Riederer, A. M., & Foster, S. O. (2009). The livelihood vulnerability index: A pragmatic approach to assessing risks from climate variability and change—a case study in Mozambique. *Global Environmental Change, 19*, 74–88.

IOM & MPI. (2014). *Integrating migration into the post-2015 United Nations development agenda* (Issue Briefs No.10). Geneva: IOM.

James, R., Otto, F., Parker, H., Boyd, E., Cornforth, R., Mitchell, D., & Allen, M. (2014). Characterizing loss and damage from climate change. *Nature Climate Change, 4*(11), 938–939. doi:10.1038/nclimate2411.

Kreutzmann, H. (1993). Socioeconomic transformation in Hunza Northern Areas, Pakistan. *Mountain Research and Development, 13*(1), 19–93.

Kreutzmann, H. (2011). Scarcity within Opulence: Water management in the Karakorum Mountain revisited. *Journal of Mountain Science, 8*, 525–534.

Laczko, F., & Collett, E. (2005). *Assessing the Tsunami's effects on migration. Feature story on migration information source.* Migration Policy Institute. http://www.migrationinformation.org/feature/display.cfm?ID=299. Accessed 16 Aug 2010.

Le De, L., Gaillard, J. C., & Friesen, W. (2013). Remittances and disasters: A review. *International Journal of Disaster Risk Reduction, 4*, 34–43.

Levitt, P. (2001). Transnational migration: Taking stock and future directions. *Global Networks, 1*(3), 195–216.

Lindley, A. (2009). *Migrant remittances in the context of crisis in Somali society. Humanitarian Policy Group Background Paper.* London: Overseas Development Institute.

Malik, A., & Piracha, M. (2006). Economic transition in Hunza and Nager valleys. In H. Kreutzmann (Ed.), *Karakoram in transition. Culture, development and ecology in the Hunza Valley* (pp. 359–369). Oxford/New York/Karachi: Oxford University Press.

Mazzucato, V., Van Den Boom, B., & Nsowah-Nuamah, N. N. (2008). Remittances in Ghana: Origin, destination and issues of measurement. *International Migration, 46*, 103–122. doi:10.1111/j.1468-2435.2008.00438.x.

Mohapatra, S., Joseph, G., & Ratha, D. (2009). *Remittances and natural disasters. Ex-post response and contribution to ex-ante preparedness* (World Bank Policy Research Paper 4972). doi:10.1596/1813-9450-4972

Mueller, V., Gray, C., & Kosec, C. (2014). Heat stress increases long-term human migration in rural Pakistan. *Nature Climate Change.* doi:10.1038/NCLIMATE2103.

Mustafa, D. (2005). The production of an urban hazardscape in Pakistan: Modernity, vulnerability and the range of choice. *The Annals of the Association of American Geographers, 95*(3), 566–586. ISSN: 0004-5608.

Myers, N. (2002). Environmental refugees: A growing phenomenon of the 21st century. *Philosophical Transactions of the Royal Society, 357*, 609–613.

Osili, U. O. (2004). Migrants and housing investments: Theory and evidence from Nigeria. *Economic Development and Cultural Change, 52*(4), 821–849.

Osili, U. O. (2007). Remittances and savings from international migration: Theory and evidence using a matched sample. *Journal of Development Economics, 83*, 446–465.

Paulson, A. (2003). *Insurance motives for migration: Evidence from Thailand. Mimeo.* Evanston: Kellogg School of Management, Northwestern University.

Pelling, M. (1998). Participation, social capital and vulnerability to urban flooding in Guyana. *Journal of International Development, 10*(4), 469–486.

Pelling, M., & High, C. (2005). Understanding adaptation: What can social capital offer assessments of adaptive capacity? *Global Environmental Change, 15*, 308–319.

Schrieder, G., & Knerr, B. (2000). Labour migration as a social security mechanism for smallholder households in Sub-Saharan Africa: The case of Cameroon. *Oxford Development Studies, 28*, 223–236. doi:10.1080/713688309.

Scoones, I. (1998). *Sustainable rural livelihoods: A framework for analysis* (Working Paper 72). Brighton: Institute for Development Studies.

Shah, K. U., Dulal, H. B., Johnson, C., & Baptiste, A. (2013). Understanding livelihood vulnerability to climate change: Applying the livelihood vulnerability index in Trinidad and Tobago. *Geoforum, 47*, 125–137.

Stark, O., & Bloom, D. E. (1985). The new economics of labor migration. *American Economic Review, 75*, 173–178.

Stark, O., & Levhari, D. (1982). On migration and risk in LDCs. *Economic Development and Cultural Change, 31*(1), 191–196.

Stark, O., & Lucas, R. (1988). Migration, remittances and the family. *Economic Development and Cultural Change, 36*(3), 465–481.

Suhrke, A. (1994). Environmental degradation and population flows. *Journal of International Affairs, 47*(2), 473–496.

Sultana, F. (2010). Living in hazardous waterscapes: Gendered vulnerabilities and experiences of floods and disasters. *Environmental Hazards, 9*(1), 43–53.

Taylor, E. J. (1999). The new economics of labour migration and the role of remittances in the migration process. *International Migration, 37*(1), 63–88. doi:10.1111/1468-2435.00066.

The World Bank. (2012). Migration and development brief, 19. The World Bank. http://siteresources.worldbank.org/INTPROSPECTS/Resources/334934-1288990760745/MigrationDevelopmentBrief19.pdf. Accessed 8 Feb 2016.

United Nations Framework Convention on Climate Change (UNFCCC). (2015). Adoption of the Paris Agreement: Proposal by the President. Draft decision -/CP.21. http://unfccc.int/resource/docs/2015/cop21/eng/l09r01.pdf. Accessed 10 Jan 2016.

Vincent, K. (2007). Uncertainty in adaptive capacity and the importance of scale. *Global Environmental Change, 17*, 12–24.

Warner, K., van der Geest, K., Kreft, S., Huq, S., Harmeling, S., Koen, K., & deSherbinin, A. (2012). *Evidence from the frontlines of climate change: Loss and damage to communities despite coping and adaptation.* Loss and damage in vulnerable countries initiative. Policy report 9. United Nations University Institute for Environment and Human Security (UNU-EHS), Bonn.

Wisener, B., & Luce, H. R. (1993). Disaster vulnerability: Scale, power and daily life. *GeoJournal, 30*(2), 127–140.

Yang, D., & Choi, H. J. (2007). Are remittances insurance? Evidence from rainfall shocks in the Philippines. *The World Bank Economic Review, 21*, 219–248.

Chapter 2
Role of Remittances in Building Farm Assets in the Flood Affected Households in Koshi Sub-Basin in Nepal

Soumyadeep Banerjee, Bandita Sijapati, Meena Poudel, Suman Bisht, and Dominic Kniveton

2.1 Introduction

Some of the most serious consequences of anthropogenic climate change are believed to be those related to changes in hydrological systems. Societies, individuals, groups, and governments are likely to adapt to future changes in climatic conditions in the same way that they have adjusted their behavior to the impacts of climate variability and extremes in the past (Adger et al. 2005; Agrawal and Perrin 2008).[1] The impacts of future climate change could be significantly reduced if people could cope better with present climate risks (Thomalla et al. 2006). Analyses of past impacts and responses to climate shocks and stressors are necessary to assess the feasibility of future responses to changing climate conditions, even if future climatic shocks and stressors are historically unprecedented (Agrawal and Perrin

---

[1] The Intergovernmental Panel on Climate Change (IPCC) defines adaptation as 'the process of adjustment to actual or expected climate and its effects. In human systems, adaptation seeks to moderate or avoid harm or exploit beneficial opportunities (IPCC 2014: 5).'

S. Banerjee (✉)
International Centre for Integrated Mountain Development (ICIMOD), Kathmandu, Nepal

University of Sussex, Sussex, UK
e-mail: soumyadeep.banerjee@icimod.org

B. Sijapati
Centre for the Study of Labour and Mobility (CESLAM), Kathmandu, Nepal

M. Poudel
International Organization for Migration (IOM), Kathmandu, Nepal

S. Bisht
International Centre for Integrated Mountain Development (ICIMOD), Kathmandu, Nepal

D. Kniveton
University of Sussex, Sussex, UK

2008). Building the adaptive capacity of individuals, groups, or organizations to adapt to changes and transforming this capacity into action are two dimensions of adaptation to a changing climate (Adger et al. 2005).[2]

Throughout history, migration has been a critical adaptation strategy to changes in natural resource condition and environmental hazards (McLeman and Smit 2006). The New Economics of Labour Migration (NELM) suggested that migration is a risk diversification strategy for rural households. In context of weakly developed credit and insurance markets, migration of a household member to seek employment provides an alternative route to reduce risk through income diversification (Stark and Bloom 1985; Taylor 1999) and improve their livelihoods (de Haas 2007). The NELM considers migration to be a household decision, wherein costs and returns of migration are shared by the migrant and non-migrating members of a household (Stark and Bloom 1985; Stark and Lucas 1988). Human mobility in response to environmental shocks or stressors could take many forms. Hugo (1996) conceptualized a mobility continuum with forced migration and voluntary migration occupying the two extremes. It is expected that the majority of climate-related human mobility will involve movements within countries with migrants using established networks and relationships to seek livelihood opportunities in response to climate change impacts (Bardsley and Hugo 2010). Social networks provide access to jobs, accommodation, and protection to the migrant workers (Tacoli 2011). Remittances – financial and social – contribute to climate change adaptation (CCA), disaster risk reduction (DRR), and development at the household level. Environmental disaster, usually, do not disrupt financial remittances, which supplement income of the recipient households (ADB 2012). There is evidence that inflow of financial remittance increases in the aftermath of environmental disasters (Yang and Choi 2007). Financial remittances could be one of the alternative financing sources that helps to manage risk from extreme events such as drought or flood. Social remittances in form of skills, information, network, and knowledge could contribute to awareness raising, income diversification, disaster risk reduction, and capacity building.

Even though remittances could increase the adaptive capacity of people who live in areas that are at high risk from frequent extreme events and compensate for the property damage, the households might not be able to avoid the recurring damage (ADB 2012). Any potential benefit from migration needs to be weighed against potential costs (e.g. social costs, unrealistic expectations, poor standard of living, and low wages or substandard working conditions in destination) (Foresight 2011). Different perceptions of the role of migration in socio-economic development, limited evidence on the relationship between migration and environmental stressors, and methodological challenges have thus resulted in a debate on whether and how environmental degradation would give rise to mass displacement and migration

---

[2] The IPCC (2014: 21) defines adaptive capacity 'as the ability to adjust, to take advantage of opportunities, or to cope with consequences'. A household could build its adaptive capacity by expanding the tangible resources used to maintain livelihoods (e.g. natural capital and productive resources) and capabilities to do so (e.g. social and human capital) (Bebbington 1999).

(Tacoli 2011), and the extent to which migration can contribute to CCA among migrant sending households and origin communities.

Mirroring the contemporary academic discourse, the migration and climate change policy discourse in the 1990s had focused on how environmental shocks and stressors would induce large-scale displacement and out-migration, identifying potential 'hot-spots', and potential destinations of these displaced populations or migrants. These early deliberations raised a specter of large-scale forced movement of people from rural to urban areas and from developing to developed countries due climate change impacts in the future (see IPCC 1990). During the past decade, there has been a shift in the dominant paradigm in migration and development discourse that returned the focus to the positive impacts of migration on origin communities due to remittances sent back by migrant workers, skills brought back by returnees, and diaspora effects on investment and support (ADB 2012). This paradigm shift has been gradually imbibed in the parallel discourse on migration and climate change. For instance, the Cancún Adaptation Framework of 2010 recognized that migration can be an adaptation strategy (ADB 2012), or by the acknowledgement in the Sendai Framework for Disaster Risk Reduction 2015–2030 that community resilience could be enhanced by reducing disaster risk through the knowledge, skills and capacities of the migrants (UNISDR 2015: 21). However, the role of human migration, particularly labor migration and remittances as a risk management strategy or CCA have received little attention in the national adaptation policies across the Hindu Kush Himalayan (HKH) region, including in Nepal. Rather migration is perceived as a challenge to development and adaptation goals. The National Adaptation Programme of Action in Nepal (NAPA) identified rural-urban migration as a challenge to urban planning process (MoE 2010: 5), and posited the need to address rural-urban migration by supporting rural development (MoE 2010: 14). This perception indicates that the context dependent nature of migration outcomes is not well understood by the national stakeholders. This is partly due to the lack of empirical evidence on the relationship between migration, CCA, and risk management.

A review of the slim evidence base on migration and environmental change in Nepal leads to the following inferences. First, environmental change is more likely to influence local rather than long-distance mobility (particularly, international migration) (see Massey et al. 2007; Bohra-Mishra and Massey 2011). Second, there is greater likelihood that a household member would migrate for work in communities exposed to rapid onset water hazard (e.g. riverine flood, flash flood) than those exposed to slow onset water hazard (e.g. dry spell, water shortage) (see Banerjee et al. 2011). Third, remittances are commonly spent on food, consumer goods, healthcare, education, and loan repayment rather than disaster risk reduction (see Banerjee et al. 2011). This review indicates towards the following gaps. Previous research had focused on small-scale case studies, which had been conducted using disparate methodologies and without standardized concepts and terminologies related to environmental stressors, CCA and migration. Most of these case studies explored the influence of environmental stressors on migration decision-making process. Thus, the evidence base regarding the role of migration and remittances in CCA, including an examination of circumstances under which financial and social remittances contribute to household level adaptive capacity, a systematic assessment of stakeholder narratives, and an assessment of the role of institutions in facilitating migration as a risk management strategy, is relatively limited.

This chapter will assess the role of financial remittances (hereafter remittances) in building farm assets (e.g. farm size, livestock, irrigation, and farm mechanization), which are an important component of a rural household's adaptive capacity. A better understanding of the determinants that shape the farm assets of a remittance recipient household is vital to understand the risk management mechanisms, which in turn would support appropriate policy initiatives. The next section provides a description of the study area in the Sagarmatha Transect of the Koshi sub-basin (KSB). This is followed by a discussion on research methodology and presentation of empirical evidence that characterizes the role of remittances in shaping farm assets among households in rural communities affected by the floods. The last section of this chapter discusses the implications of this research.

2.2 Case Study

This case study is part of the Rural Livelihoods and Climate Change Adaptation in the Himalayas programme ('the Himalica') of the International Centre for Integrated Mountain Development (ICIMOD). The KSB is located in Nepal's eastern Ganges region. The catchment area of this sub-basin is composed of the mountain region in the north, through the mid-hills, to *Terai* region (plains) in the south. The KSB is known for floods and extremely high sediment load (Dixit et al. 2009). The Himalica programme selected the Sagarmatha Transect within the KSB through stakeholder consultations that included government and non-government institutions. This transect includes the mountain district of Solukhumbu, mid-hill districts of Khotang and Udayapur, and Terai district of Saptari (see Fig. 2.1).

The KSB, which is comprised of the eastern highland, lowlands of the Ganges, and one of three snow-fed watersheds in Nepal, provides a unique research context. This region is known for the impacts of rapidly changing ecosystems, shifts in hydrological patterns, changes in land use and concomitant pressures on ecosystems and livelihoods. This region experiences recurrent extreme events such as cloudbursts, flash floods and droughts. In future, the frequency and severity of these extreme events are expected to increase due to climate change (NCVST 2009; Dixit et al. 2009). For the purposes of the study, riverine and flash floods are used as a proxy for future climatic change induced extreme events. The KSB is historically known for high mobility of able-bodied men, initially to serve in the British and Indian armies, and later on a seasonal basis to India in search of better livelihood opportunities. This trend has continued to the present day. An increasing number of people are participating in circular labor migration to the middle-east and south-east

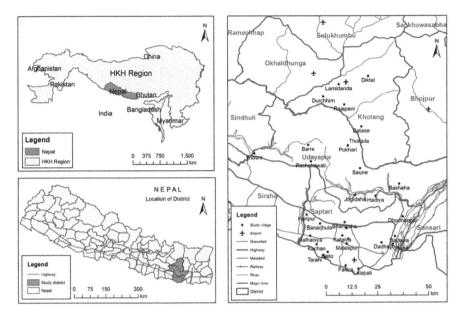

Fig. 2.1 Map of the study area in the Sagarmatha Transect, Koshi sub-basin, Nepal (Source: Migration Case Study, Himalica programme, ICIMOD)

Asia. Further information on demographic, socio-economic, floods and migration characteristics in the four districts of the Sagarmatha Transect in the KSB is provided in Table 2.1.

2.3 Research Methodology and Methods

Agriculture in the study area is mainly subsistence in nature, and is combined with other economic activities. Agricultural land and livestock are important components of a rural household's adaptive capacity (Aulong et al. 2012) and represent an accumulation of wealth (Vincent 2007). The farm size and number of livestock owned by a household are attributes of a household's farm assets. The application of tools, implements and powered machinery to enhance agricultural production and productivity and reduce drudgery is referred as 'mechanization'. (Clarke 2000). Farm mechanization could lead to an increase in crop productivity, address labour shortage, or support a change in cropping pattern. In this study, farm mechanization includes use of tractors to plough the farm during the winter cropping season or ownership of tractor, power tiller, or mechanized threshers. Access to irrigated farm land could reduce environmental risk by reducing dependence on rain-fed agriculture. The use of irrigation during the winter cropping season indicates flexibility with in a household's farming portfolio.

Table 2.1 Brief description of the study area in the Sagarmatha Transect, Koshi sub-basin, Nepal

District	Arable land		Flood affected VDCs		Population		Absentee population			HDI	Per capita income ($PPP)
	Hectares	Percentage	Number	Percentage	Number	Household	Male	Female	Total		
Solukhumbu	15,546.2	4.69	7	20.59	105,886	23,785	4,730	887	5,617	0.502	1,841
Khotang	26,077.9	16.39	51	44.74	206,312	42,664	16,504	1,158	17,662	0.494	1,132
Udayapur	25,794.9	12.5	17	22.37	317,532	66,557	20,036	2,024	22,060	0.475	920
Saptari	67,054.5	49.2	18	40.91	639,284	121,098	25,676	752	26,433	0.437	801

Source: DWIDP (2009, 2010, 2012), CBS (2013, 2012), UNDP (2014)

To study the relationship between remittances and farm assets, a separate regression was conducted for each of the aforementioned attributes of farm assets. Migration is considered to be a risk-sharing behavior of the household to diversify resources in order to minimize income risks (Stark and Levhari 1982). Remittances are sent by migrant workers in destination to their families in the origin community.[3] The remittance recipient status of the household (i.e. remittance recipient or non-recipient) is the indicator of mobility and one of the independent variables. To quantify the marginal effect of remittances, a number of other independent variables were taken into account: household head's age, gender, ethnicity, and educational attainment; household's dependency ratio, flood damage (i.e., financial losses due to flood) to the household between 1994 and 2013; time taken to reach nearest paved road, local market, and bank; village level meetings on flood preparedness; and adjusted monthly per capita expenditure. A modified version of the same regression model was used to characterize the farm assets of the remittance recipient households in the study area. The household survey had recorded the duration for which a household had been receiving remittance, which is the period between the first and latest instances of remittance receipt by a household. It was recorded as a continuous variable in the household survey, and was converted into a categorical variable: short duration (i.e. below median value) and long duration (i.e. above median value). The methodology for this part of the chapter draws from Banerjee et al. (forthcoming).

This research study adopted a mixed method approach that included focus group discussions (FGD), household and village surveys, and key informant interviews (KII). The fieldwork was conducted in 2014.

Initially, the FGDs were conducted in all four districts of the Sagarmatha Transect: Solukhumbu, Khotang, Udayapur, and Saptari districts. Since floods are not a major environmental stressor in Solukhumbu district, the survey was conducted only in the remaining three districts, which were considered as one aggregated areal unit. A list of all flood affected rural wards was prepared and the selection of households involved a two stage process.[4] In the first stage, the Probability Proportional to Size (PPS) was used to select rural wards. In the second stage, equal number of households were selected using systematic sampling within each selected rural ward. A household was classified as either a remittance 'recipient' household or 'non-recipient' household.[5] At the end of the survey, a sample size of 333 was achieved; 159 remittance recipient households and 174 non-recipient households.

---

[3] In this study, a household was considered to be a migrant-sending household if any household member had lived and worked in another village or town in the same country or another continuously for two months or more at any time during the last 30 years. Households not conforming to this definition were considered as non-migrant households.

[4] If a village development committee (VDC) had experienced a riverine flood or flash flood at least once since 1984 then all rural wards in the VDC was considered as flood affected.

[5] If at any time during the past 30 years a household had received financial remittances, irrespective of the relationship of the remittance sender to the household, it was referred as a remittance recipient household.

2.4 Results

2.4.1 Household Responses to Floods

The household level responses to floods in the study area have been distinguished between the responses during the flood (or inundated) period, immediate aftermath of the flood, and between two flood events. There is little difference in the flood response strategies of the remittance recipient and non-recipient households in the flood affected rural communities. The common household level responses during the flood are shifting of cattle and family to a safer location, storing valuables in a safe place within the house, buying food on credit, reducing the proportion or number of meals, borrowing money from relatives or friends, relying on less preferred food, digging a ditch or channel to divert flood water, and spending savings to procure food. In the immediate aftermath of a flood, household level responses are focused on recovery measures such as repairing the house or cattle-shed, borrowing money from friends or relatives, buying food on credit, spending savings to procure food, relying on less preferred food, preparing for farming, reducing proportion or number of meals, and bringing back cattle from safe location. Likewise, household measures between two flood events are raising the plinth of the house or cattle-shed or granary, building a barrier to reduce the speed of flood water, repairing local infrastructure (e.g. bridge, road), reducing area under paddy, and borrowing money from relatives or friends. It is evident that the household responses during the flood and in its immediate aftermath in study area are short-term in nature, and even strategies adopted to reduce impact over the long-run are quite rudimentary. The reasons for this are many but primarily could be attributed to inability of households to divert resources to build adaptive capacities; lack of support from government and non-government agencies; limited access to information and technology; and lack of collective will. During a focus group discussion in Udayapur, a female member recounted,

> We are aware of the fact that we have to do something to about the damages caused by the flood. Every year, we incur heavy losses … … We have to build an embankment. But what can we do? If we do not work, the stoves in our houses will not light. Everyone is concerned about their own needs because we all are poor. This is also why we have not been able to do anything.

Another explained,

> It is mostly because of insufficient investment. It is difficult to prevent flood damages at the household level because until and unless anything is done at the village level, that is, prevent the river from flowing into the village, preventive measures at the household level will not be sufficient … … There are embankments that are constructed but with low investments so they are not strong and easily get destroyed … … We hardly receive any support from the government and lack of co-ordination and collective spirit among the villagers is also another major issue.

The village level flood preparedness is limited. Only 20% of the households reported that their village has a community level flood contingency plan. The availability of a pre-designated flood shelter for the villagers and livestock was reported by 6% of the households. However, half of households reported to have participated in a village meeting that had discussed flood preparedness.

2.4.2 Migration and Rural Livelihoods

As mentioned earlier, circular migration in search for employment and higher earnings has for long been a defining feature of the livelihoods of many households in the study area. Migrant workers from the villages studied are predominantly men of working age group. Factors such as decline in agricultural productivity because of floods, inability of youths with modicum levels of education (i.e., those who have failed their Grade 10 exams or passed it but not pursued higher education) to be gainfully employed, lack of income-generating opportunities in the villages or surrounding localities, and prospects of economic improvement through remittances are common reasons for migration.

With regards to migrant destinations, out of the 656 migration episodes since 1984, over half are associated with a destination in a third country such as Malaysia, Qatar, Saudi Arabia, or United Arab Emirates.[6] Another quarter of the destinations are located within Nepal, and the remaining are destinations in India. Most of the migrant workers are wage employees (95%) in the destination communities who are employed in the secondary sector, primarily, in construction (35%) and manufacturing (18%).[7] Most of these migrant workers to do not receive social security benefits (e.g. pension, provident fund, and insurance) from their employers. For instance, only 15% of the surveyed migrant workers had access to some form of social security benefits as part of their latest job. Only 14% of the surveyed migrant workers are entitled to paid leave.

Remittances are an important component of the recipient household's income (see Fig. 2.2). Around two-third of the remittance recipient households have identified remittances as the major source of household income. The mean volume of remittances received from domestic and foreign sources during the 12 months preceding the survey is estimated to be USD 243 and USD 1,112 respectively.[8] The mean remittance per capita for the remittance recipient household is estimated to be USD 294. On an average, recipient households have received remittances for a period of 5.5 years. Notably, compared to the non-recipient households, the remit-

---

[6] International migration for work to countries other than India.

[7] Other minor employers included electric, gas, and water supply, defence services, and hotels and restaurants.

[8] Exchange rate in December 2015 was 1 USD = NPR 106.40

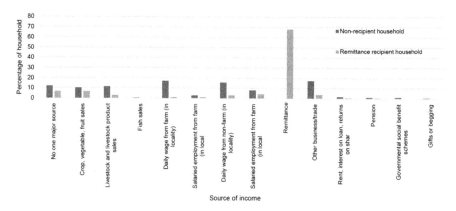

Fig. 2.2 Major source of household income during the 12 months preceding the survey, Sagarmatha Transect, Koshi sub-basin, Nepal, 2014 (Source: Migration Case Study, Himalica programme, ICIMOD)

tance recipient households are less dependent on other sources of income in the origin community (e.g., wage employment, business/trade, and agricultural activities). Thus indicating that in most instances, remittances serve as the main source of a household's income, rather than supplementing that from other sources.

2.4.3 Remittances and Adaptive Capacity

Remittances are, commonly, used to procure food and consumer goods, repay loans, and access healthcare and education. Some households have invested remittances in agricultural input and communication (see Fig. 2.3). There was little use of remittances in housing, savings, community activities, disaster risk reduction, insurance, or business input. When asked about the reasons for such low expenditure on disaster risk reduction, the common refrain was that individual households have to budget for various requirements of their household members. The disaster preparedness measures are generally considered to be of low priority, especially when confronted with daily necessities such as food, education, loan repayment, and health care. As recounted by a male migrant returnee from Khotang,

> There have been slight changes after migration. I have not been able to buy land or build a house from the income I earned abroad. But what the remittances did help in is in sustaining household expenditures and repaying loans that I had taken to go abroad… There is now nothing left to invest on anything [including preparedness measures to overcome the impact of floods]

Among the various attributes of farm assets, the regression analysis found a significant association between remittances and large farm size ('above median size'),

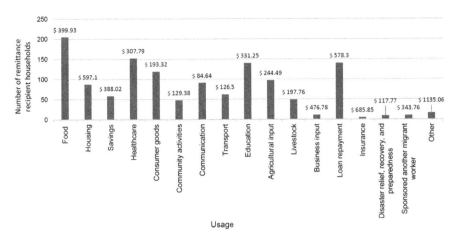

Fig. 2.3 Use of the financial remittances during the 12 months preceding the survey, Sagarmatha Transect, Koshi Sub-basin, Nepal, 2014 (Source: Migration Case Study, Himalica programme, ICIMOD)

Table 2.2 Effects of remittances on household level farm assets in Sagarmatha Transect, Koshi sub-basin, Nepal

	Recipient	Non-recipient	Beta coefficient
% of households with farm size above the median value[a]	61.4	49.6	0.497*
% of households with livestock above the median value[b]	61.2	67.3	−0.094
% of households with irrigated farm above the median value[c]	61.4	49.6	−0.141
% of households that had used irrigation during the winter cropping season	7.9	3.6	0.698
Farm mechanization	12.9	10.1	0.285

Source: Migration Case Study, Himalica programme, ICIMOD
Note: Non-recipient households were considered as referent. All regressions models were household head's age, gender, ethnicity, dependency, and literacy; expenditure category; flood damage to the household between 1994 and 2013; time to reach nearest paved road, local market, bank; and participation in village level meeting on flood preparedness was organized
* $p<0.1$; ** $p<0.05$; *** $p<0.01$.
[a]Estimated only for households with access to farm land
[b]Estimated only for households with access to livestock
[c]Median value for farm size = 0.41, +Median value for livestock = 5

which is an important component of rural households' adaptive capacity (see Table 2.2). More precisely, remittance recipient households are likely to have access to larger farms than non-recipient households. This finding should be considered in context of another finding that longer the duration for which remittance is received by a household, less likely it was to have a farm above the median size (see Table 2.3).

Table 2.3 Effect of duration of remittance receipt on household level farm assets among the remittance recipient households in Sagarmatha Transect, Koshi sub-basin, Nepal

	Shorter duration	Longer duration	Beta coefficient
% of households with farm size above the median value[a]	61.4	52.9	−0.926**
% of households with livestock above the median value[b]	50.0	52.7	−0.063
% of households with irrigated farm above the median value[c]	44.3	48.5	−0.028

Source: Migration Case Study, Himalica programme, ICIMOD
Note: Shorter duration remittance receipt households were considered as referent. All regression models were adjusted for the household head's age, gender, ethnicity, dependency, and literacy; expenditure category; flood damage to the household between 1994 and 2013; time to reach nearest paved road, local market, bank; and participation in village level meeting on flood preparedness was organized
* $p<0.1$; ** $p<0.05$; *** $p<0.01$.
[a]Estimated only for households with access to farm land
[b]Estimated only for households with access to livestock
[c]Median value for farm size=0.41, +Median value for livestock=5

Thus, of the various forms of farm assets such as land, livestock, irrigation, and farm mechanization, that have bearings on households' adaptive capacity, the only one that is positively associated with remittances is farm size. Though, over the long term, there was a decrease in farm size of a remittance recipient household. This is an important finding since agriculture is not only one of the major livelihood strategies in the study area; agricultural productivity is not only volatile but also confronts risk from environmental stressors.

2.5 Discussion

Labor migration is an important livelihood strategy of many households in the study area, and remittances are a major source of income for remittance recipient households. In the study locations remittances are commonly spent on food, healthcare, loan repayment, education, and consumer goods. There is little targeted savings to manage flood risk, investment of remittances in insurance, or other measures pertaining to disaster risk reduction. The common household responses during the flood and in its immediate aftermath are reactive and short-term in nature and those between two flood events include some low-cost structural measures. This study observed a significant positive association between remittance recipient status of a household and access to farm above median size. Farm size is an important component of rural households' adaptive capacity. However, findings from the study also suggest that over the long term a remittance recipient household is less likely to have farm size above the median size. This indicates that the remittance recipient households in the study area are likely to downsize their farm-holding. Presumably, they become dependent on remittances as a livelihood strategy rather than using it as a once off attempt at wealth accumulation. The vagaries of weather, price, and crop diseases create risks for farming (Lucas 2014). Smaller farm size among long duration remittance recipient households may indicate towards a downsizing of farming activities and growing dependence on remittances and on the local market for food and other commodities. Based on a study in China, Tao and Xu (2007) had suggested that unlike the older and the less educated laborers in rural area, young educated migrants would not value farming as much. They would tend to disassociate themselves from farming in future. If given an opportunity to migrate permanently, they might even de-link themselves from the farm land allocated to them.

These results have important implications for the study area. On one hand, subsistence farming is still a major component of livelihoods and food security of many households. At the same time, the number of Nepali migrant workers has been progressively increasing over the past couple of decades. The findings from regression analyses suggest that in the long-term a sizeable proportion of remittance recipient households are likely to reduce their farm size, and in turn their dependence on subsistence agriculture would decline. Given the environmental sensitivity of subsistence farming in general, and particularly due to the lack of irrigation, farming technology, and farm mechanization, a reduction in farm size could reduce the environmental risks posed on the household's livelihoods portfolio. On the other hand, this would also imply a possible rise in the household's dependence on the market to procure food items, which is likely to affect long-term food security, particularly of women, young children and the elderly. At present, Nepal imports several major food items (e.g. rice, vegetables, cooking oil, fruits) from neighboring India. A disruption of supplies from across the border could further reduce food security of the households. Farm sector provides earning opportunities in the form of self-employment as well as wage-employment to many in the study area. A reduction in farm size is likely to leave these people without any sources of income. This trend poses a challenge to meet Nepal's national priority goals since agriculture is a priority sector, and the government programs aim to increase productivity and diversify agriculture sectors to address food insecurity and nutritional status.[9]

The Local Adaptation Plan for Action (LAPA) exercise at the VDC level provides a flexible framework to address local risks as well as take advantage of opportunities (GoN 2011). The role of migration and remittances has to be mainstreamed in the LAPA exercises. A coordinated awareness-raising program is required for relevant stakeholders such as functionaries of the District Development Committee (DDC) and Village Development Committee (VDC), remittance recipient households, development partners, and financial institutions. For example, the LAPA

---

[9] 13th Periodical Plan 2013–2016, source: http://www.npc.gov.np/images/download/13th-Plan.pdf, accesses Nov 2015.

exercise needs to recognize that remittances could increase purchasing power of the remittance recipient households that can develop the adaptive capacity of households by building farm assets, and also raise the food security (both quantity and quality) of the remittance recipient households. At the same time, this exercise should also identify a wide array of food sources to strengthen the food supply and avoid excessive dependence on any particular source.

Remittances could support off-farm livelihood diversification, especially in the form of cash crops and off-farm wage employment, which are priority sectors for national planning processes, and can in turn compensate for the income losses in the farm sector during the lean season or extreme events. Equally important would be to enhance market linkages, support capacity building of remittance recipient households, providing them with access to credit for supporting cash crop and off-farm employment. As it is the women who are left behind in the remittance receiving households gender-sensitive planning and support provisions are required.

These measures will require programmatic commitment in line with policy priority, as well as guidance, nurturing, and technical inputs from relevant government and non-government agencies. Steps should be taken to diversify the skill sets of the farm labor, particularly because of the positive association observed between remittances and farm size. It would not only contribute to a possible rise in income, but will also support diversification of household livelihoods portfolio. Furthermore efforts should be made, possibly with matching funds from the government and non-government sources, to incubate and nurture small and medium rural enterprises established by the remittance recipients.

The contribution of remittance in the development finance of Nepal has not been acknowledged. The contribution of international remittance is phenomenal – foreign aid represents 15 % of national budget in Nepal and the amount of remittance Nepal received is roughly equivalent to the size of the government's annual budget for the year 2014/2015.[10] This is equivalent to around 29 % of the country's GDP. Nepal, thus, stands third in the world and top in South Asia in terms of the contribution of remittance in GDP. In addition to keeping a stable source of foreign exchange earnings that helps keep balance of payment afloat, migrants usually send more money when the family back home experiences hardships such as disasters. Therefore remittances act as an insurance against economic adversity which has been well exemplified during the recent April 29, 2015 earthquake. While these examples clearly reveal the significance of remittances, this reality seems to be largely ignored by the national planning processes. Nepal needs to acknowledge the role of remittances into development and adaptation processes and move towards a comprehensive migration policy that addresses challenges and opportunities provided by remittances, including at the local and household levels.

---

[10] Nepal Rastra Bank calculates the remittance in the tune of Rs 600.17 billion for the year 2014/015 through the formal banking channel alone. Labour Migration for Employment: A Status Report for Nepal 2014/2015, Ministry of Labour and Employment, Government of Nepal (forthcoming January 2016).

Acknowledgement The authors express their gratitude to the editors for providing this opportunity to prepare this chapter. Particular appreciation goes to Dr. Koko Warner (UNU-EHS) and Dr. Andrea Milan (UNU-EHS) for their support, encouragement and insightful feedback. The authors would like to thank Dr. Golam Rasul (ICIMOD) and Mr. Valdemar Holmgren (ICIMOD) for their support and encouragement. Special thanks to Dr. Bidhubhusan Mahapatra (ICIMOD) for his guidance with the statistical analysis. Particular appreciation also goes to the CESLAM team who had provided invaluable support during the fieldwork. The authors would like to thank Dr. Vishwas Chitale (ICIMOD) and Mr. Gauri Dangol (ICIMOD) for their inputs in the preparation of the maps and graphs. The authors will like to thank the anonymous reviewers for their helpful comments. This research was supported by the Rural Livelihoods and Climate Change Adaptation in the Himalayas (Himalica) Programme, which is implemented by the International Centre for Integrated Mountain Development (ICIMOD), and is funded by the European Union.

References

Adger, W. N., Arnell, N. W., & Tompkins, E. L. (2005). Successful adaptation to climate change across scales. *Global Environmental Change, 15*, 77–86(2000).
Agrawal, A., & Perrin, N. (2008). *Climate adaptation, local institutions, and rural livelihoods* (IFRI Working Paper # W08I-6). Michigan: International Forestry Resources and Institutions Program. School of Natural Resources and Environment. University of Michigan.
Asian Development Bank. (2012). *Addressing climate change and migration in Asia and the Pacific*. Mandaluyong City: Asian Development Bank.
Aulong, S., Chaudhurim, B., Farnier, L., Galab, S., Guerrin, J., Himanshu, H., & Reddy, P. P. (2012). Are South Indian farmers adaptable to global change? A case in Andhra Pradesh catchment basin. *Regional Environmental Change, 12*, 423–436.
Banerjee, S., Gerlitz, J. Y., & Hoermann, B. (2011). *Labour migration as a response strategy to water hazards in the Hindu Kush-Himalayas*. Kathmandu: ICIMOD.
Banerjee, S., Kniveton, D., Black, R., Bisht, S., Das, P. J., Mahapatra, B., & Tuladhar, S. (forthcoming). *Do financial remittances build household level adaptive capacity? A case study of the flood affected households of Upper Assam in India* (KNOMAD Working Paper).
Bardsley, D., & Hugo, G. (2010). Migration and climate change: Examining thresholds of change to guide effective adaptation decision-making. *Population and Environment, 32*, 238–262.
Bebbington, A. (1999). Capitals and capabilities: A framework for analyzing peasant viability, rural livelihoods and poverty. *World Development, 27*(12), 2021–2044.
Bohra-Mishra, P., & Massey, D. (2011). Environmental degradation and out-migration: New evidence from Nepal. In E. Piguet, A. Pécoud, & P. de Guchteneire (Eds.), *Migration and climate change* (pp. 74–101). Cambridge: Cambridge University Press.
Central Bureau of Statistics. (2012). *National population and housing census 2011*. Kathmandu: National Planning Commission Secretariat, Government of Nepal.
Central Bureau of Statistics. (2013). *National sample census of agriculture Nepal 2011/12* (District Summary). Kathmandu: National Planning Commission Secretariat, Government of Nepal.
Clarke, L. J. (2000). *Strategies for agricultural mechanization development the roles of the private sector and the Government*. Rome: FAO. Italy: Agricultural Support Systems.
de Haas, H. (2007). *Remittances, migration and social development: A conceptual review of the literature* (Social Policy and Development Programme Number 34). Geneva: United Nations Research Institute for Social Development.

Department of Water Induced Disaster Prevention. (2009). *Disaster review (2009)*. Kathmandu: Ministry of Irrigation, Government of Nepal.
Department of Water Induced Disaster Prevention. (2010). *Disaster review (2010)*. Kathmandu: Ministry of Irrigation, Government of Nepal.
Department of Water Induced Disaster Prevention. (2012). *Disaster review (2012)*. Kathmandu: Ministry of Irrigation, Government of Nepal.
Dixit, A., Upadhya, M., Dixit, K., Pokhrel, A., & Rai, D. R. (2009). *Living with water stress in the hills of the Koshi Basin, Nepal*. Kathmandu: ICIMOD.
Foresight. (2011). *Migration and global environmental change: Final project report*. London: The Government Office for Science.
Government of Nepal. (2011). *National framework on local adaptation plans for action*. Singhdurbar: Government of Nepal, Ministry of Environment.
Hugo, G. J. (1996). Environmental concerns and international migration. *International Migration Review, 30*(1), 105–131.
Intergovernmental Panel on Climate Change IPCC. (1990). Policymakers' summary of the potential impacts of climate change (Report from Working Group II to IPCC) IPCC – Secretariat, Geneva.
IPCC (Intergovernmental Panel on Climate Change). (2014). *WGII AR5 glossary. Working group II contribution to the intergovernmental panel on climate change fourth assessment report*. Cambridge: Cambridge University Press.
Lucas, R. E. B. (2014). Internal migration in developing economies: An overview. Paper produced for KNOMAD's Thematic Working Group (TWG) on Internal Migration and Urbanization. KNOMAD.
Massey, D., Axinn, W., & Ghimire, D. (2007). *Environmental change and out-migration: Evidence from Nepal, population studies center research report 07-615*. Ann Arbor: Population Studies Center.
McLeman, R., & Smit, B. (2006). Migration as an adaptation to climate change. *Climatic Change, 76*, 31–53.
Ministry of Environment. (2010). National adaptation programme of action to climate change, Kathmandu, Nepal
Nepal Climate Vulnerability Study Team (NCVST). (2009). *Vulnerability through the eyes of vulnerable: Climate change induced uncertainties and Nepal's development predicaments*. Kathmandu: ISET-N. Boulder, (Colorado).
Stark, O., & Bloom, D. E. (1985). The new economics of labor migration. *American Economic Review, 75*, 173–178.
Stark, O., & Levhari, D. (1982). On migration and risk in LDCs. *Economic Development and Cultural Change, 31*(1), 191–196.
Stark, O., & Lucas, R. (1988). Migration, remittances and the family. *Economic Development and Cultural Change, 36*(3), 465–481.
Tacoli, C. (2011). Migration and mobility in a changing climate: A policy perspective. *Revista Interdisciplinar da Mobilidade Humana Brasília, Ano XIX, 36*, 113–124.
Tao, R., & Xu, Z. (2007). Urbanization, rural land system and social security for migrants in China. *The Journal of Development Studies, 43*(7), 1301–1320. doi:10.1080/00220380701526659.
Taylor, E. J. (1999). The new economics of labour migration and the role of remittances in the migration process. *International Migration, 37*(1), 63–88. doi:10.1111/1468-2435.00066.
Thomalla, F., Downing, T., Spanger-Siegfried, E., Han, G., & Rockström, J. (2006). Reducing hazard vulnerability: Towards a common approach between disaster risk reduction and climate adaptation. *Disasters, 30*(1), 39–48.
UNDP. (2014). *Nepal human development report 2014: Beyond geography, unlocking human potential*. Kathmandu: United National Development Programme and National Planning Commission Secretariat, Government of Nepal.

UNISDR. (2015). Sendai framework for disaster risk reduction 2015–2030. http://www.wcdrr.org/uploads/Sendai_Framework_for_Disaster_Risk_Reduction_2015-2030.pdf. Accessed 8 Feb 2016.

Vincent, K. (2007). Uncertainty in adaptive capacity and the importance of scale. *Global Environmental Change, 17*, 12–24.

Yang, D., & Choi, H. (2007). Are remittances insurance? Evidence from rainfall shocks in the Philippines. *The World Bank Economic Review, 21*(2), 219–248.

Chapter 3
Migration as a Risk Management Strategy in the Context of Climate Change: Evidence from the Bolivian Andes

Regine Brandt, Raoul Kaenzig, and Susanne Lachmuth

3.1 Introduction

The effects of global climate change are likely to increasingly impact social and environmental systems (IPCC 2014). In the tropics, recent warming trends have been greatest at higher altitudes, drawing attention to possible future negative effects on mountain regions and water resources (Kohler et al. 2014). In the Central Andes, recent warming trends are considered a possible factor behind the acceleration of glacier retreats (Francou et al. 2000; Rabatel et al. 2013; Soruco et al. 2009; Vuille et al. 2003), a phenomenon which could significantly reduce glacial water buffers. There are already shortages in water supply, especially during the dry seasons (Bradley et al. 2006). Further climate-change-related stressors that are likely to have

---

R. Brandt (✉)
Formerly International Network on Climate Change (INCA), Institute of International Forestry and Forest Products, Technical University Dresden,
Pienner Str. 7, 01737 Tharandt, Germany

Institute of Biology/Geobotany and Botanical Garden, Martin-Luther-University Halle-Wittenberg, Am Kirchtor 1, 06108 Halle/Saale, Germany
e-mail: regine.brandt@botanik.uni-halle.de

R. Kaenzig
Institute of Geography, University of Neuchâtel,
Espace Louis-Agassiz 1, 2000 Neuchâtel, Switzerland
e-mail: raoul.kaenzig@unine.ch

S. Lachmuth
Institute of Biology/Geobotany and Botanical Garden, Martin-Luther-University Halle-Wittenberg, Am Kirchtor 1, 06108 Halle/Saale, Germany

German Centre for Integrative Biodiversity Research (iDiv) Halle-Jena-Leipzig,
Deutscher Platz 5e, 04103 Leipzig, Germany
e-mail: susanne.lachmuth@botanik.uni-halle.de

a negative impact on Central Andean livelihoods include the shifting precipitation patterns and the increasing intensity and frequency of weather extremes (e.g. heavy rainfall, frost, and intense heat). Peasant farming systems are particularly vulnerable and face increasing threats because they are highly dependent on climate-sensitive water, soil, and biodiversity (Perez et al. 2010; Valdivia et al. 2010). Additional risk factors for peasant livelihoods are the remoteness from economic and political centers and the limited access to basic services (e.g. education, health care, and infrastructure). These conditions foster rural poverty in mountain areas (Jodha 2005; McDowell and Hess 2012), because they restrict the range of livelihood response options for households to cope with and adapt to climate-related stressors (Morton 2007; Stadel 2008).

Households commonly seek to provide for the needs of their members against disruptions coming from the climate, market economy or other factors (Netting 1993). A frequent mountain household response aimed at reducing livelihood insecurity and social vulnerability is rural out-migration (Grau and Aide 2007). From among different theories of migration, the New Economics of Labor Migration (NELM) (Stark and Bloom 1985) provides a useful frame for empirical work on climate-related migration (see Kniveton et al. 2008). Households are theorized in NELM as the relevant unit of analysis for understanding the economic strategies and migratory decisions of individual members. From this perspective, migration is a strategy for income diversification and risk management, which commonly aims at improving a household's absolute and relative socioeconomic status within their respective community (Stark and Bloom 1985).

There are multiple drivers behind the decisions to migrate (De Haas 2010; Massey et al. 1993; Piguet 2012). The role of environmental stressors, such as those related to climate change, has been widely debated, especially during the last two decades. The first alarmist prognostics about "climate refugees" (e.g. Myers 1993) have been counterbalanced by studies that underline the multiplicity of factors in migratory processes (Black 2001; Castles 2002). Thus, environmental factors are most likely intertwined with social, economic, demographic, and political drivers (Black et al. 2011; De Sherbinin et al. 2011; Kniveton et al. 2008; Perch-Nielsen et al. 2008; Piguet 2013; Warner 2010).

In the Central Andes, the effects of environmental drivers on migratory movements were studied. In the southern Ecuadorian Andes, for instance, Gray (2009) discovered that environmental factors were strongly related to short-distance migration but not to international migration. In the central highlands of Peru, rainfall variability and climatic conditions also played an important role in rural farmers' migration decisions, which were, however, primarily shaped by economic factors unrelated to the climate (Milan and Ho 2014). In the field of environment-related migration, recent investigations emphasize the importance of studying migrants' perceptions of environmental changes and risks (Piguet 2010), because they influence individuals' expectation of being exposed to threats and their subsequent strategies to cope with these risks (Grothmann and Patt 2005). In order to understand how households adapt to environmental changes, considering migration as a possible response, it seems essential to document these changes and to assess how they are perceived by the people who are exposed and need to adapt to them either by undertaking migration themselves or by influencing their household members' migration decisions.

In the Bolivian Andes, few empirical case studies have so far tackled the linkages between environmental changes and migration as a related risk management strategy (e.g. Balderrama et al. 2011; Guilbert 2005; Kaenzig 2014; McDowell and Hess 2012; Zoomers 2012). In this region, mobility is a traditional livelihood strategy in response to both environmental and social risks. Harsh climatic conditions in the diverse mountain environments, in particular, have led Andean households to cope with stressors through constant relocations in order to diversify resources and the risks associated with their procurement (McDowell and Hess 2012; Valdivia et al. 1996). Broader migratory movements from the rural highlands to rural lowlands, urban areas, and destinations abroad, have been increasingly incorporated into the activities of Bolivian Andean households, particularly since the neoliberal shifts in economy and politics in the 1980s (Sanabria 1993). The "New Economic Policy", implemented in 1985, aimed to reduce the Bolivian state's deficits and to stimulate economic growth at the national level, yet it led to increased living costs and massive unemployment rates, especially in the mining sector. The policy supported agricultural imports from abroad and agro-industrial development solely in the lowlands, thus negatively affecting the peasant economies of the highlands (Sanabria 1993). A prolonged drought in the middle of the 1980s, combined with the shifts in economic policy presented above, additionally contributed to an accentuated impoverishment of the rural inhabitants of the Bolivian Andes (Balderrama et al. 2011). As a consequence, peasants and also miners and unemployed urban workers massively migrated to the coca production sites in *Yungas* (department of La Paz) and *Chapare* (department of Cochabamba) (Sanabria 1993), both of which are transition areas between the Andean highlands (~4200 masl) and the Amazon (~500 masl) along the eastern slopes of the Bolivian Andes (Navarro and Maldonado 2002). Overall migration to these areas declined in the 1990s when anti-coca policies were implemented and coca was mainly substituted by less profitable cash crops (Bradley and Millington 2008).

Like other Andean regions, the northern *altiplano* (high Andean plateau) and the high Andean valleys of the department of La Paz, where research for this study was conducted, are also characterized by high migration rates, mostly from rural to urban areas (Balderrama et al. 2011; Cortes 2000; Guilbert 2005; Mazurek 2007; Zoomers 2012). Since the 1970s, for instance, the city of El Alto has been a preferred destination for rural migrants (Lazar 2008). In the last decades, the urban center has grown from a small town to a city with an estimated population of almost one million (INE 2015). Around 75 % of the city's inhabitants identify themselves as belonging to the Aymara ethnolinguistic group (INE-PNUD 2005), which is a clear indication of their rural origin.

In this chapter, we focus on extending the analysis of migration patterns and the drivers behind the decisions to migrate from rural areas of the Bolivian Andes to consider migration as a strategy to cope with the risks related to climate change. We investigated two case study areas located in the rural municipality of Palca, situated in the high Andean valleys close to the city of La Paz. In our analysis, we explored the migration patterns and the drivers of migratory movements and, based on these data, we analyzed whether environmental factors, and especially those related to climate change, played a crucial role in influencing the decisions to migrate.

3.2 Methods

3.2.1 Research Area

In 2013, the present research was conducted in the rural municipality of Palca, which is situated in the province of Murillo in the department of La Paz in Bolivia. The population of Palca consists of almost 16,000 residents (INE 2015). Aymara is the main spoken language, and around half of the inhabitants are bilingual with Spanish as the second language. Almost 70 % of the population relies on small-scale rain-fed and irrigated farming for their subsistence, which is complemented by temporary and permanent off-farm activities (e.g. mining, manufacture) in the locality, nearby urban centers or in the lowlands. Around 80 % of the population live in extreme poverty (INE-PNUD 2005).

The municipality is located in a high Andean valley region at an altitude between 3000 and more than 6000 m above sea level, which receives an average of 557 mm of annual precipitation, with more than 70 % of the rainfall occurring between November and March. The annual mean temperature in the municipal capital Palca (3400 masl) is 12 °C and varies between 5 and 22 °C (CATIE-OTN Bolivia 2012).

We conducted one case study in six communities belonging to the municipality within the Choquekota river basin, which receives its water from the Mururata glacier (5880 masl) (area: *Mururata*), and a second case study in four communities within the Sajhuaya river basin located close to the Illimani glacier (6439 masl) (area: *Illimani*) (Fig. 3.1). During the last decades, both glaciers have seriously retreated: Mururata has lost 20 % of its ice area (1956–2008; Ramírez 2009) and Illimani 35 % (1963–2009; Da Rocha Ribeiro et al. 2013).

Due to the Illimani ice loss, one of its nearby communities, Khapi, has become an important symbol of climate change in the international media (Kaenzig 2014). The notoriety of the Illimani glacier as a symbol of the effects of climate change played an important role in determining the selection of this geographic area for our research. Additionally, the presence of important urban centers attracting rural migrants also played a role in deciding to conduct our study here.

3.2.2 Data Collection

In each of the two watersheds indicated above, 68 local migrant households (with household heads reporting about current migrants and with returnees talking about their past migration experiences) were randomly selected and analyzed (Illimani

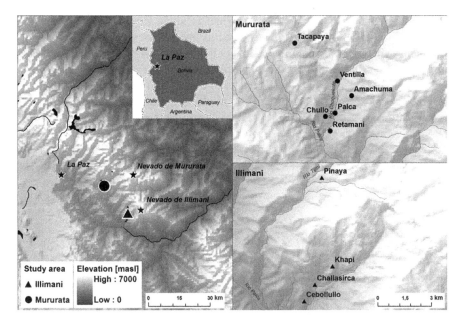

Fig. 3.1 The research was conducted in the municipality of Palca in the department of La Paz in Bolivia in six communities that receive water from the Mururata glacier (area: Mururata; marked with *dots*) and in four communities located close to the Illimani glacier (area: Illimani; marked with *triangles*) (Source: Own data collected in Illimani and Mururata, Bolivian Andes, in 2013)

n=41, Mururata n=27). Based on prior informed consents and mutual written agreements with at least one respondent per household (confidential data use, data storage and processing in anonymized form, scientific data use only), the data were collected by using semi-structured interviews (Atteslander 2003) to explore the characteristics of the household members and their migration events. The interviews included "freelisting exercises" (Quinlan 2005) for identifying the most salient drivers for migration. The freelisting method, in which it is assumed that the respondents mention important aspects more frequently and earlier than others, enables easily applicable and efficient data collection (e.g. Quinlan 2005). Freelisting has been used as a data collection method in ethnobiological studies (e.g. Brandt et al. 2013) and is also applied in migration studies related to, for instance, health (Maupin et al. 2011). It can be easily combined with other common methods used in climate-related research, such as the participatory rural appraisal (PRA) (Kartiki 2011; Meze-Hausken 2000; Rademacher-Schulz et al. 2012). Languages used for the interviews were either Spanish or Aymara, in the latter case with the support of a local interpreter.

In both case studies, the interviews focused on the destinations and frequency of migration events for the members of each selected household. In our inquiry, migration events were all the temporary (temporally restricted or circular) geographical movements of one or more months per year to places outside the Palca municipality by household members who had a residence both within and outside the study areas. Migration events referred to the interviewees' entire lives and also included the permanent movements of individuals related to the house-holding nuclear family, who had their main residence outside the area of origin but who contributed regularly or occasionally to the livelihood of their household of origin in Palca through remittances, products or other kinds of support. Furthermore, we collected data about the factors driving migration for the members with migration experience (past or present) of each selected household. In both areas, the drivers mentioned were recorded at the household level. In Mururata, they were additionally recorded for each specific migration event of the household members. In the latter research area, we also noted the sex and level of education of the migrants.

3.2.3 Data Analysis

The drivers for migration reported within each household (in both study areas) and for each migration event per household (in Mururata only) were listed. Their relative importance was calculated based on their frequency of citation (FC) (Quinlan 2005) without considering how early they were cited, because not all data used were collected in a way that allowed us to determine the migration drivers' order of mentioning. They were mainly categorized into economic (eco), environmental (env) and human-socio-cultural (soc) factors. A low number of citations were relegated to the joint residual category of political and demographic (pol) factors; all categories according to Black et al. (2011) (modified). We performed generalized linear mixed model analyses (GLMM, R package "lme4", function "glmer") in R software for statistical computing and graphics; version 3.1.2 for Windows (R Development Core Team 2015). The FC of drivers for migration within categories (eco, env, soc, pol) was used as response variable (binary data with binomial error distribution). All maximal models were fit with a maximum likelihood approach and simplified in a stepwise-backward procedure based on likelihood ratio tests (LRT, χ^2) (Crawley 2007). In detail, we first assessed whether the FC of drivers within each driver category mentioned by the participants in our research differed significantly between both study areas. To avoid pseudo-replication, we took into account a random effect of the community to which each household belonged. Secondly, we analyzed whether the FC of drivers within each category mentioned for each migration event per household (n = 50; in Mururata only; no data available for Illimani) varied depending on the respective migrants' characteristics (sex, school education) and destinations. We took into account nested random effects of the household and community related to the migrants. Subsequently, Tukey post-hoc tests were applied for multiple comparisons of group means (Crawley 2007).

3.3 Results

3.3.1 Migration Dynamics

In the studied rural households, temporary and permanent migration activities were shown to be key elements of livelihood strategies. The 68 households interviewed were, on average, composed of approximately five members. Most commonly, the households consisted of married couples living together with their children. Grandparents or adult brothers and sisters also belonged to half of the households in our sample. In total, we documented 163 temporary and permanent migration events (Illimani n=99; Mururata n=64), thus an average of 3 migration events per household.

The most popular migrant destinations found in our study were the urban areas of La Paz and El Alto (64%). Altogether, regional movements within the department of La Paz represented more than 80% of the 163 migration events that we recorded. Forty percent of the migration events were temporary (temporally restricted or circular), meaning that the respective regional migrants had double residency and maintained the farming activities in their areas of origin while working or studying in the nearby urban area. Often the most productive household members migrated. We documented, for instance, that 94% of the migrants from Mururata were between 14 and 38 years old (female: 64%; male: 36%). Within the selected households of our study areas only few migration events were directed towards the formerly popular (before the 1990s) rural and nearby migration destination *Yungas* (8%) at the time of or recently before our field work. Also, very few migrants chose urban destinations in other departments of Bolivia, such as Cochabamba (2% of the migration events) or Santa Cruz (3%). Eight percent of the migration events recorded were towards the neighboring countries Argentina (3%) and Brazil (4%), or to Europe (Italy and Spain; 1%).

3.3.2 Multiple Drivers for Migration

The lack of work opportunities and income, the search for higher education opportunities, the scarcity of farming land and water, and the desire to pursue an urban life style, were the predominant drivers for migration mentioned by our study participants (Fig. 3.2). While all international migration events were at least partly driven by economic factors, these were, however, significantly less mentioned as drivers for movements at the regional (58% of migration events) or national level (30%). In contrast, at the regional and national levels, not only economic but multiple drivers play an important role for migration, such as, the search for education opportunities (Table 3.1).

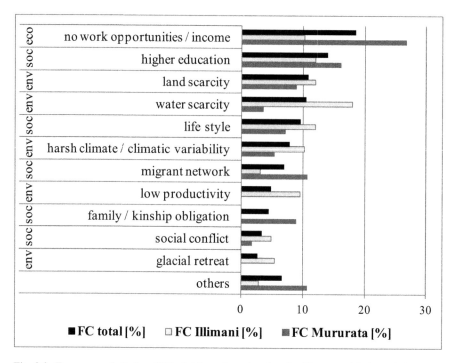

Fig. 3.2 Frequency of citation (FC) of drivers for migration (in %) reported in households from Illimani (n=27) and Mururata (n=41) in La Paz, Bolivia, and categorized into economic (eco), environmental (env) and human-socio-cultural (soc) factors (Source: Own data collected in Illimani and Mururata, Bolivian Andes, in 2013)

Table 3.1 Effects of the migrants' sex, education and destination on the frequency of citation (FC) of economic (eco), environmental (env), human-social-cultural (soc) and political-demographic (pol) drivers for migration in the study area Mururata

Fixed effect	FC							
	eco		env		soc		pol	
	χ^2	p	χ^2	p	χ^2	p	χ^2	p
Sex	0.01	ns	0.67	ns	2.78	ns	7.01	**
Education	4.60	ns	4.53	*	4.10	*	2.56	ns
Destination	17.42	***	14.50	***	0.65	ns	5.72	ns

Source: Own data collected in Illimani and Mururata, Bolivian Andes, in 2013
ns not significant; results based on the minimal adequate generalized linear mixed models (GLMM) for the FC of drivers for migration as response variables; significance of fixed effects determined by likelihood ratio tests LRT, Chi square, χ^2); levels of significance: *, $p<0.05$; **, $p<0.01$; ***, $p<0.001$

In more than 60% of the regional movements, the search for higher education drove the migration decision, while for long distance movements this factor was almost not mentioned. Thus, higher education (e.g. technical school, university) as a driver for migration was significantly related to the destination of migration.

Mentioning this driver was also significantly dependent on the migrants' school qualification, as high-school graduation is a prerequisite for higher education. In contrast, for movements of individuals with low or intermediate school qualifications, the search for better education was almost never mentioned in the interviews ($< 3\%$).

3.3.3 Migration and Climate Change

In our study areas, almost all of the respondents pointed to signs of increasing climatic variability, which can be assumed to be a result of climate change, expressed by rising unpredictability of rainfall and extreme weather events, such as hail and frost, as well as increasing temperatures. Between both study areas, the FC of drivers for migration differed significantly ($\chi^2_{(1)} = 16.86$, $p < 0.001$) with regard to environmental factors, such as water scarcity or climatic factors, with higher predicted mean values in Illimani (97.6%) than in Mururata (22.2%). The mentioning of environmental drivers also differed with the educational level and in relation to the migrants' destinations (Table 3.1), but the differences between the means were not statistically significant.

3.4 Discussion

3.4.1 Migration Dynamics

In our case studies, migration activities were shown to be key elements of rural peasants' livelihood strategies. They were mostly directed to the nearby urban areas of La Paz and El Alto. Studies from other rural highland areas with high out-migration rates, such as in northern Potosí, indicate similar trends as in our study region, where movements at a regional level were most often found (Balderrama et al. 2011). According to Mazurek (2007), internal migration in Bolivia commonly takes place within the same region, especially towards nearby cities, and is often linked with multi-residency and commuting between urban and rural areas, emphasizing the migrants' close socio-economic and cultural ties with their areas of origin (Mazurek 2007). These links may serve as a "safety net" (Zoomers 2012: 125). In comparison, the low number of migrants that were directed towards the nearby rural *Yungas* is indicative of the developments of these areas since the implementation of anti-coca policies in the 1990s, when smallholder farming became less profitable in this region.

In accordance with the national level patterns outlined by Mazurek (2007), very few migrants went to urban destinations in other departments of the country. In Bolivia, migration flows at the national level commonly take place from and towards the cities, whereas long distance rural out-migration is less frequent (Mazurek 2007). In line with these general observations, movements from rural areas to international destinations were also rare in our sample. The low frequencies of international migration can be explained by the fact that rural migrants first need to accumulate economic capital in the cities in order to be able to handle the higher costs and risks of longer distance migration (Balderrama et al. 2011). Furthermore, social capital, such as migrant networks, plays a crucial role in long distance movements. Especially with international migration, networks between migrants and non-migrants linked through ties of kinship, friendship, and similar origin, decrease the costs and risks of migration, while increasing the expected outcomes (Massey et al. 1993). The importance of social capital for international migration was emphasized in our study area Mururata, where our respondents indicated migrant networks as important factors for migration in around 80% of the international movements that we documented. The low rates of international movements recorded in our study area are likely to persist also in the future, given the local scarcity of relevant social capital.

3.4.2 Multiple Drivers for Migration

Migration is commonly driven by the desire to increase and diversify income in order to manage the potential risks of household maintenance. In this section, we discuss the main drivers for temporary and permanent migration activities documented in our study area; these include the lack of work opportunities and income, the search for higher education opportunities, the scarcity of farming land and water, and the desire to pursue an urban life style – independently and in relation to some characteristics of the respective migrants.

Our results on migration drivers reflect the high socio-economic pressure on rural Andean households dependent on both subsistence farming and off-farm work. Off-farm work has become an important source for peasant household income and is, most often, associated with temporary or permanent migration (McDowell and Hess 2012; Zoomers 2012), because in rural areas off-farm work opportunities are usually rare. In Mururata, for instance, they are almost exclusively found in the gold mining sector. Our respondents of Mururata indicated the search for jobs and income sources (economic factors) as drivers for movements at any distance. According to other studies conducted in the region, the fact that often the most productive household members commute or temporarily migrate can increase the strain on the remaining persons, mostly women, elderly and children who have to substitute the lost labor power in the households (De la Riva et al. 2013; McDowell and Hess 2012). The results, indicating that the economic drivers for migration varied significantly in relation to migrants' destinations (see Table 3.1), can be explained by the fact that international migration is primarily undertaken by locals with the perspective of earning money for themselves and their families. In contrast, as our results show, multiple drivers, such as the search for education opportunities play an important role for migration at the regional and national level but not at the international level. This is understandable given that education requires significant financial investment before any returns can be expected.

The fact that migration is determined by multiple factors is most visible in the case of young persons. They are very likely to migrate, given that they perceive the scarcity of off-farm jobs and low agricultural production and thus meager incomes as major obstacles to the satisfaction of their increasing livelihood needs (see also Balderrama et al. 2011), especially during the life stage of family formation. In our study area, many young people migrated to look for off-farm sources of income, because their access to land was very limited. As a 38 year-old migrant explained: "*The land is divided for all brothers and sisters. […] Thus, we only have small plots. […] We also have children, and if we divide our small-sized plots for all of them, they won't have any chance to produce anything. That is why people are leaving.*" What this migrant pointed to was the practice of *minifundio* (land division between heirs), applied since the Bolivian Agrarian Reform Law in 1952, when the hacienda system was dismantled, and the farming land was redistributed among the communities of highland indigenous people (Bottazzi and Rist 2012; Kay and Urioste 2007). This practice, which results in land fragmentation, combined with the basic scarcity of adequate agricultural land in the tight valleys with steep slopes where our research sites were located, and with limited options for mechanization, explains why agricultural production alone was often not sufficient to cover basic living costs. Moreover, the living costs were increasingly high due to the integration of rural areas in the highly monetized global market economy. It is thus understandable as to why, for most of the young people in our study area, agriculture was at most only a source of additional income.

Our results are consistent with the findings of McDowell and Hess (2012) who conducted research in the same region. These authors also reported that scarcity of farming land, income and rural off-farm work opportunities, as well as the increasing need for financial capital, all push the rural inhabitants to migrate, especially to the nearby urban centers. Still, despite these trends, subsistence agriculture is highly valued and maintained for ensuring food security, which was also observed in Peru (e.g. Milan and Ho 2014) and Ecuador (e.g. Gray and Bilsborrow 2014; Jokisch 2002). The results of Gray and Bilsborrow (2014) even suggest an expansion of cultivated land with an increasing number of migration events, while agricultural activities are reduced (lower labor input and production per unit area). In the Bolivian departments Chuquisaca and Potosí, income gained from migratory activities is also commonly not invested in intensifying agricultural production in the home communities, but rather used for buying urban plots or facilitating higher education for children (Zoomers 2012). Similar patterns were observed in some provinces in Ecuador, where agricultural subsistence production remains an important activity of migrant households, but given the economic and agricultural limitations, an investment in agricultural cultivation is rarely undertaken and rural landscapes gradually transform into peri-urban landscapes (Jokisch 2002). In Palca, however, off-farm income was observed to be used also for covering the costs for intensified agriculture (McDowell and Hess 2012), which often includes the use of artificial fertilizers and pesticides in order to improve agricultural productivity and compensate for low or shifting prices (e.g. De la Riva et al. 2013). This has gained momentum, also shown in our case studies. Though, despite how popular the practices are in general, they were by no means endorsed by all peasants. Thus, some of our respondents also underlined the negative effects of using artificial fertilizers like, for instance, soil compaction as well as the low quality and deficient taste of the agricultural products.

Migration is not only a strategy for increasing household income in the short term, but most likely also a strategy to attain a higher level of education which, in turn, enables one to enhance income and social status in the long term. In order to prepare themselves for finding a niche in urban job markets, many young people from the rural areas seek higher education in the cities, which results in movements to the neighboring urban areas where universities and schools for professional qualification are located. In this context, La Paz and El Alto were the most important destinations for the young inhabitants of our research areas, who were in many cases driven by the search for higher education; especially migrants who had previously graduated from high school. This means that the better qualified people (in terms of education) migrated for increasing their educational level, while for those with little education it most likely did not make sense to pursue higher education in the city before exhausting all local opportunities. In general, the frequency of citation of socio-cultural factors as drivers for migration increased in importance proportionally with the educational level of the respective migrants (Table 3.1).

Education-driven migration is, in most cases, likely to not benefit the areas of origin of the migrants. As other authors have shown, rural out-migration and the integration into urban life often imply changes in migrants' lifestyle and young migrants usually do not go back to their rural areas of origin (Balderrama et al. 2011). In our study, it was also shown that the search for an urban lifestyle was itself an important driver for migration (see Fig. 3.2). Thus, it is very likely that those rural migrants, who live in a city for extended periods of time, turn from temporary rural-urban migrants to permanent urban settlers. As a consequence, the communities of the area of origin are often depleted of the migrants' social and cultural capital (Balderrama et al. 2011). However, the increasing rural-urban networks in the surroundings of La Paz and El Alto also allow for returns of economic, social and cultural capital to rural areas like our study region. There is a need for further studies on the role of migrants' remittances on the socio-economic and demographic development in the rural areas of La Paz and other regions of the Central Andes.

3.4.3 Migration and Climate Change

Our discussion of the research results concerning the relation between climate change and migration follows two complementary lines of analysis. Firstly, we discuss the relative importance of climatic and environmental events attributable to climate change as factors motivating the respondents to migrate. Secondly, we shed light on the relations that can be drawn between climate change and other factors that push people from our study area to migrate.

A direct connection between climate change and migration in the discourse of our informants was more visible in the case of our study region Illimani. Our respondents in Illimani emphasized more climatic variability, glacier retreat and associated water problems as drivers of migration than our study participants of Mururata (Fig. 3.2). The Illimani glacier has retreated more than the Mururata glacier (see Da Rocha Ribeiro et al. 2013; Ramírez 2009) and has also become in recent years a public symbol of glacial retreat caused by climate change in Bolivia (see Kaenzig 2014). Against this background, we can interpret the documented differences between our two case study areas as being shaped both by objective factors (actual glacier retreat) and by wider discourses. In order to locate the role of climate change in migration, we consider it thus necessary to discuss not only the direct link drawn by our informants between climatic factors and migration decisions, but also to understand the role played by climate change in influencing a complex of factors that affect Andean livelihoods.

To isolate analytically the role of climate change in shaping the livelihoods of Andean rural inhabitants and their migratory movements is a difficult task, as illustrated by our results concerning the mentioning of water scarcity as a factor driving migration, which was indicated by our respondents in both study areas. The mentioning of water scarcity is understandable given that it directly affects agricultural production and food security in areas such as the semi-arid Bolivian Andes (see also Zoomers 2012). But to what extent is the water scarcity indicated by our informants a driver of migration related to climate change? Sketching an answer to this question allows us to shed light on the broader issue of the influence of climate change on the complex of factors impacting on the security and welfare of Andean livelihoods and, consequently, on migration as a strategy to manage risks and redress insecurities.

A direct and immediate connection between climate change and migration by way of its direct effects on water scarcity is difficult to make. In our informants' accounts, water availability problems were related to the intensification of agriculture, technical defects of irrigation systems, lack of water governance and management regulations. The salience of such explanations can be on the one hand related to the urgency of solving everyday intra- and inter-communal conflicts related to inequalities in access to water. But it may also be that such factors affecting access to water are genuinely more important at the moment, given that glacier retreat does not immediately translate into a relevant influence on the quantity of water runoff (see Ramirez 2009). Moreover, despite the current glacier retreats caused by atmospheric warming, decreased precipitation rates and reductions in cloud covers, both glaciers in our research areas, Illimani and Mururata, are not expected to disappear completely during the next decades (Ramirez 2009; De Rocha Ribeiro et al. 2013). While our informants who were living close to Mururata were far from being alarmed by glacier retreat as a factor behind water scarcity, they were aware of declines in the quality of water, especially during summer, when ice melting shows its maximum peaks. In relation to these results, we suppose that the influence of climate change on the availability of water will likely be more salient and also visible for local inhabitants as water runoff will decrease in the long term (see Kaenzig 2014).

The discussion regarding water scarcity indicates that the effects of climate change are intertwined with other factors affecting the livelihoods of Andean rural residents, most importantly those factors affecting the satisfaction of basic subsistence needs through agricultural production. Thus, the salience of climate change as a factor influencing migration has to be understood not simply in relation to the mentioning of climatic factors by respondents but, even more importantly, in relation to the salience of economic factors in pushing people away from their areas of residence. *"The agricultural production is put at risk here, because of the hails and frosts episodes"* told a 30 year-old migrant woman. *"You can make money, but you can also lose everything. It's a big risk! We do not have a fixed salary, like monthly or weekly. It is just based on the production. In La Paz you receive your money every two weeks or every month. This is why we go there."* Visible in this quotation is the interconnection between climatic events, the insecurity of agricultural production and of livelihoods, and migration. Balderrama et al. (2011) have indicated that climate change started to manifest in this area with a prolonged drought in the 1980s followed by alterations in rainfall patterns. As these authors pointed out, the peasants who remained involved in farming in the highlands have changed some of their traditional agricultural practices in their attempt to adapt to a delay of the planting season, a reduction in the number of days with rainfall and the occurrence of dry periods even in the rainy season, which all decreased the agricultural productivity. Visible in our study is that migration is an additional strategy of coping with risks related to climate change and less secure agricultural production. Surely, the effects of climate change on migration are not straightforward, but mediated by rural Andean inhabitants' limited abilities of adapting their production to and coping with the risks of the changed environmental conditions and by the degree of their dependence on agricultural production.

3.5 Conclusion

Our study aimed to provide a better understanding of the migration patterns in the Bolivian Andes and the drivers behind the migrants' decisions in the context of risk management, focusing on the cases of two high Andean valleys close to the city of La Paz. The methodological approach adopted in this paper allowed for exploring the multiple drivers shaping migration decisions in the Bolivian Andes and for determining their relative importance. The perspective that we adopted in this research also focused on the inter-linkages between the drivers, while also keeping in consideration the wider narratives of the interviewees.

Our results can be summarized as follows. Firstly, temporary and permanent migration from rural areas is a traditional household strategy in the Bolivian Andes, which has, from our interpretation of the interview data, increased in importance as a consequence of rising economic pressures, scarcity of land and off-farm work opportunities and low agricultural productivity in the areas of origin of the migrants. Secondly, rural out-migration is predominantly undertaken by young women and men who want to increase their incomes and educational level. Migration mostly takes place within the same geographical and cultural region and is often directed to nearby urban areas, which are, in our study area, La Paz and El Alto. Thirdly, the climatic variability, which has increased as a result of climate change, plays a role as an additional stressor for agricultural production, as well as water scarcity. The latter is, however, not only connected to lower water runoff as an effect of climate change and increasing glacier melting, but also to water overuse, technical irrigation problems, social conflicts, and the lack of water governance. Both climatic variability and water scarcity alone may not directly drive migration, but they are usually combined with other stressors impacting agricultural production and livelihood security.

In the face of climate change, there is, on the one hand, an urgent need for more rigorous support to the rural peasant households in coping with the agricultural and economic risks related to increasing climatic variability, water scarcity and the consequences of warming trends, not only in the Bolivian and Central Andes but also in other mountainous regions in the world. Especially the salient influence of social factors on water availability indicates that making more rigorous political decisions on local, regional, and national levels in order to regulate water management and to decrease the pressure on scarce water resources is very important. A lack of involvement here may otherwise lead to increasing social conflicts that undermine the social cohesion of the region, not only in the rural areas but also in the urban ones. On the other hand, it is also of crucial importance to develop rural off-farm work opportunities and improve basic services, such as schools and infrastructure, in order to help the rural peasant youth to diversify and improve their income sources in both rural and urban areas. For achieving these tasks, more research on the drivers and dynamics of rural migration can provide crucial background information.

Acknowledgement Funding was granted by the DAAD (Deutscher Akademischer Austauschdienst), Bonn, Germany. We would like to express our sincere gratitude to all interviewees who shared their experience and time with us. We also acknowledge the cooperating Instituto de Ecología (Universidad Mayor de San Andrés, La Paz) and the project PRAA ("Adaptación al Impacto del Retroceso Acelerado de Glaciares en los Andes Tropicales", Ministerio de Medio Ambiente y Agua, Bolivia). In addition, we are very grateful to the staff members of CARE Bolivia for providing field work logistics, and to AguaSustentable, BMI (Bolivian Mountain Institute) and CATIE (Centro Agronómico Tropical de Investigación y Enseñanza) for their scientific input. Our sincere thanks also go to André Lindner, Verónica Agner, Vladimir Mendieta and Gunnar Seidler for their support in coordination, interview translation, transcription and preparation of the map, as well as to Mihai Popa for his helpful comments on the manuscript. Finally, we would like to thank the reviewers who contributed to improving this chapter.

References

Atteslander, P. (2003). *Methoden der empirischen Sozialforschung*. Berlin: Walter de Gruyter.

Balderrama, M. C., Tassi, N., Rubena, M. A., Aramayo, C. L., & Cazorla, I. (2011). *Rural migration in Bolivia: The impact of climate change, economic crisis and state policy*. Human Settlements Group. London: IIED.

Black, R. (2001). Environmental refugees: Myth or reality? *New Issues in Refugee Research (UNHCR)*, 34.

Black, R., Adger, W. N., Arrnell, N. W., Dercon, S., Geddes, A., & Thomas, D. S. G. (2011). The effect of environmental change on human migration. *Global Environmental Change, 21S*, S3–S11. doi:10.1016/j.gloenvcha.2011.10.001.

Bottazzi, P., & Rist, S. (2012). Changing land rights means changing society: The sociopolitical effects of agrarian reforms under the Government of Evo Morales. *Journal of Agrarian Change, 12*(4), 528–551.

Bradley, A. V., & Millington, A. C. (2008). Coca and colonists: Quantifying and explaining forest clearance under coca and anti-narcotics policy regimes. *Ecology and Society, 13*(1), 31.

Bradley, R. S., Vuille, M., Diaz, H. F., & Vergara, W. (2006). Threats to water supplies in the tropical Andes. *Science, 312*(5781), 1755–1756. doi:10.1126/science.1128087.

Brandt, R., Mathez-Stiefel, S.-L., Lachmuth, S., Hensen, I., & Rist, S. (2013). Knowledge and valuation of Andean agroforestry species: The role of sex, age, and migration among members of a rural community in Bolivia. *Journal of Ethnobiology and Ethnomedicine, 9*, 83. doi:10.1186/1746-4269-9-83.

Castles, S. (2002). Environmental change and forced migration: Making sense of the debate. *New Issues in Refugee Research (UNHCR)*, 70.

CATIE-OTN Bolivia (2012). *Plan de gestión integral de cuenca. Para el área de intervención del proyecto PRAA*. Programa estratégico: cuencas y cambio climático. La Paz.

Crawley, M. J. (2007). *The R book*. Chichester: Wiley.

Da Rocha Ribeiro, R., Ramírez, E., Simões, J., & Machaca, A. (2013). 46 years of environmental records from the Nevado Illimani glacier group, Bolivia, using digital photogrammetry. *Annals of Glaciology, 54*(63), 272–278. doi:10.3189/2013AoG63A494.

De Haas, H. (2010). Migration and development: A theoretical perspective. *International Migration Review, 44*(1), 227–264.

De la Riva, M. V., Lindner, A., & Pretzsch, J. (2013). Assessing adaptation – Climate change and indigenous livelihood in the Andes of Bolivia. *Journal of Agriculture and Rural Development in the Tropics and Subtropics (JARTS), 114*(2), 109–122.

De Sherbinin, A., Castro, M., Gemenne, F., Cernea, M. M., Adamo, S., Fearnside, P. M., et al. (2011). Preparing for resettlement associated with climate change. *Science, 334*(6055), 456–457. doi:10.1126/science.1208821.

Francou, B., Ramirez, E., Cáceres, B., & Mendoza, J. (2000). Glacier evolution in the tropical Andes during the last decades of the 20th century: Chacaltaya, Bolivia, and Antizana, Ecuador. *AMBIO, 29*(7), 416–422. doi:10.1579/0044-7447-29.7.416.

Grau, H. R., & Aide, T. M. (2007). Are rural-urban migration and sustainable development compatible in mountain systems? *Mountain Research and Development, 27*(2), 119–123. doi:10.1659/mrd.0906.

Gray, C. (2009). Environment, land, and rural out-migration in the southern Ecuadorian Andes. *World Development, 37*(2), 457–468. doi:10.1016/j.worlddev.2008.05.004.

Gray, C. L., & Bilsborrow, R. E. (2014). Consequences of out-migration for land use in rural Ecuador. *Land Use Policy, 36*, 182–191. doi:10.1016/j.landusepol.2013.07.006.

Grothmann, T., & Patt, A. (2005). Adaptive capacity and human cognition: The process of individual adaptation to climate change. *Global Environmental Change, 15*(3), 199–213. doi:10.1016/j.gloenvcha.2005.01.002.

Guilbert, M.L. (2005). Environnement et migration: les difficultés d'une communauté rurale andine (El Terrado, Potosi, Bolivie). *VertigO - la revue électronique en sciences de l'environnement, 6*(3). doi:10.4000/vertigo.2441

Instituto Nacional de Estadística (INE) (2015). *Estadísticas demográficas. La Paz: Población total proyectada, por sexo, según provincia y sección de provincia, 2009–2011.* Instituto Nacional de Estadística. Estado Plurinacional de Bolivia. Retrieved March 5, 2015 from http://www.ine.gob.bo.

Instituto Nacional de Estadística (INE) - Programa de las Naciones Unidas para el Desarrollo (PNUD) (2005). *Bolivia: Atlas estadístico de municipios.* Instituto Nacional de Estadística, Estado Plurinacional de Bolivia. Programa de las Naciones Unidas para el Desarrollo. Retrieved March 5, 2015 from http://www.ine.gob.bo.

Intergovernmental Panel on Climate Change (IPCC) (2014). *Climate change. Synthesis report. Summary for policymakers.* Geneva: World Meteorological Organization/Intergovernmental Panel on Climate Change.

Jodha, N. S. (2005). Adaptation strategies against growing environmental and social vulnerabilities in mountain areas. *Himalayan Journal of Sciences, 3*(5), 33–42.

Jokisch, B. D. (2002). Migration and agricultural change: The case of smallholder agriculture in highland Ecuador. *Human Ecology, 30*(4), 523–550.

Kaenzig, R. (2014). Can glacial retreat lead to migration? A critical discussion of the impact of glacier shrinkage upon population mobility in the Bolivian Andes. *Population and Environment, 34*(4), 480–496. doi:10.1007/s11111-014-0226-z.

Kartiki, K. (2011). Climate change and migration: A case study from rural Bangladesh. *Gender & Development, 19*(1), 23–38. doi:10.1080/13552074.2011.554017.

Kay, C., & Urioste, M. (2007). Bolivia's unfinished agrarian reform: Rural poverty and development policies. In A. H. Akram-Lodhi, S. M. Borras, & C. Kay (Eds.), *Land, poverty and livelihoods in an era of globalization: Perspectives from developing and transition countries* (pp. 41–79). New York: Routledge.

Kniveton, D., Schmidt-Verkerk, K., Smith, C., & Black, R. (2008). *Climate change and migration: Improving methodologies to estimate flows.* Geneva: IOM (International Organization for Migration).

Kohler, T., Wehrli, A., & Jurek, M. (2014). *Mountains and climate change: A global concern. Sustainable Mountain Development Series.* Bern: Centre for Development and Environment (CDE), Swiss Agency for Development and Cooperation (SDC), Geographica Bernensia.

Lazar, S. (2008). *El Alto, rebel city: Self and citizenship in Andean Bolivia.* Durham: Duke University Press.

Massey, D. S., Arango, J., Hugo, G., Hugo, G., Kouaouci, A., Pellegrino, A., et al. (1993). Theories of international migration: A review and appraisal. *Population and Development Review, 19*(3), 431–466.

Maupin, J. N., Ross, N., & Timura, C. A. (2011). Gendered experiences of migration and conceptual knowledge of illness. *Journal of Immigrant and Minority Health, 13*(3), 600–608. doi:10.1007/s10903-010-9333-9.

Mazurek, H. (2007). Three pre-concepts regarding the internal migration in Bolivia. *Revista de Humanidades y Ciencias Sociales (Santa Cruz de la Sierra), 14*(1–2), 1–18.

McDowell, J. Z., & Hess, J. J. (2012). Accessing adaptation: Multiple stressors on livelihoods in the Bolivian highlands under a changing climate. *Global Environmental Change, 22*(2), 342–352. doi:10.1016/j.gloenvcha.2011.11.002.

Meze-Hausken, E. (2000). Migration caused by climate change: How vulnerable are people in dryland areas? *Mitigation and Adaptation Strategies for Global Change, 5*(4), 379–406. doi:10.1023/A:1026570529614.

Milan, A., & Ho, R. (2014). Livelihood and migration patterns at different altitudes in the Central Highlands of Peru. *Climate and Development, 6*(1), 69–76. doi:10.1080/17565529.2013.826127.

Morton, J. F. (2007). The impact of climate change on smallholder and subsistence agriculture. *Proceedings of the National Academy of Sciences of the United States of America, 104*(50), 19680–19685. doi:10.1073/pnas.0701855104.

Myers, N. (1993). Environmental refugees in a globally warmed world. *BioScience, 43*(11), 752–761.

Navarro, G., & Maldonado, M. (2002). *Geografía ecológica de Bolivia: vegetación y ambientes acuáticos*. Santa Cruz: Centro de Ecología Difusión Simón I. Patiño.

Netting, R. M. (1993). *Smallholders, householders: Farm families and the ecology of intensive, sustainable agriculture*. Stanford: Stanford University Press.

Perch-Nielsen, S. L., Bättig, M. B., & Imboden, D. (2008). Exploring the link between climate change and migration. *Climatic Change, 91*(3–4), 375–393. doi:10.1007/s10584-008-9416-y.

Perez, C., Nicklin, C., Dangles, O., Vanek, S., Sherwood, S., Halloy, S., et al. (2010). Climate change in the high Andes: Implications and adaptation strategies for small-scale farmers. *The International Journal of Environmental, Cultural, Economic and Social Sustainability, 6*(5), 71–88.

Piguet, E. (2010). Linking climate change, environmental degradation and migration: A methodological overview. *Wiley Interdisciplinary Reviews: Climate Change, 1*(4), 517–524. doi:10.1002/wcc.54.

Piguet, E. (2012). Migration: The drivers of human migration. *Nature Climate Change, 2*(6), 400–401. doi:10.1038/nclimate1559.

Piguet, E. (2013). From "primitive migration" to "climate refugees": The curious fate of the natural environment in migration studies. *Annals of the Association of American Geographers, 103*(1), 148–162. doi:10.1080/00045608.2012.696233.

Quinlan, M. (2005). Considerations for collecting freelists in the field: Examples from ethnobotany. *Field Methods, 17*(3), 219–234. doi:10.1177/1525822X05277460.

R Development Core Team (2015). The R Project for Statistical Computing. R for Windows 3.1.2. Retrieved 19 Feb 2015 from http://www.r-project.org.

Rabatel, A., Francou, B., Soruco, A., Gomez, J., Cáceres, B., Ceballos, J. L., et al. (2013). Current state of glaciers in the tropical Andes: A multi-century perspective on glacier evolution and climate change. *The Cryosphere, 7*, 81–102. doi:10.5194/tc-7-81-2013.

Rademacher-Schulz, C., Tamer, A., Warner, K., Rosenfeld, T., Milan, A., Etzold, B., et al. (2012). *Rainfall variability, food security and human mobility: An approach for generating empirical evidence* (Intersections, Vol. 10). Bonn: UNU-EHS.

Ramírez, E. (2009). *Estudio de la disponibilidad de recursos hídricos de la cuenca del Nevado Mururata (Informe final)*. La Paz.

Sanabria, H. (1993). *The coca boom and rural social change in Bolivia*. Ann Arbor: University of Michigan Press.

Soruco, A., Vincent, C., Francou, B., & Gonzalez, J. F. (2009). Glacier decline between 1963 and 2006 in the Cordillera Real, Bolivia. *Geophysical Research Letters, 36*, L03502. doi:10.1029/2008GL036238.

Stadel, C. H. (2008). Vulnerability, resilience and adaptation: Rural development in the tropical Andes. *Pirineos, 163*, 15–36.

Stark, O., & Bloom, D. E. (1985). The new economics of labor migration. *The American Economic Review, 75*(2), 173–178.

Valdivia, C., Dunn, E. G., & Jetté, C. (1996). Diversification as a risk management strategy in an Andean agropastoral community. *American Journal of Agricultural Economics, 78*(5), 1329–1334.

Valdivia, C., Seth, A., Gilles, J. L., García, M., Jiménez, E., Cusicanqui, J., et al. (2010). Adapting to climate change in Andean ecosystems: Landscapes, capitals, and perceptions shaping rural livelihood strategies and linking knowledge systems. *Annals of the Association of American Geographers, 100*(4), 818–834. doi:10.1080/00045608.2010.500198.

Vuille, M., Bradley, R. S., Werner, M., & Keimig, F. (2003). 20th century climate change in the tropical Andes: Observations and model results. *Climatic Change, 59*(1–2), 75–99. doi:10.1023/A:1024406427519.

Warner, K. (2010). Global environmental change and migration: Governance challenges. *Global Environmental Change, 20*(3), 402–413. doi:10.1016/j.gloenvcha.2009.12.001.

Zoomers, A. (2012). Migration as a failure to adapt? How Andean people cope with environmental restrictions and climate variability. *Global Environment, 9*, 104–129.

Chapter 4
Circular Migration and Local Adaptation in the Mountainous Community of Las Palomas (Mexico)

Noemi Cascone, Ana Elisa Peña del Valle Isla, and Andrea Milan

4.1 Introduction

Building on a broad academic consensus, the Fourth Assessment Report of the Intergovernmental Panel on Climate Change (IPCC) showed that human activity has a significant negative impact on the climate, which in turn adversely affects populations living in vulnerable environments. Climate change is projected to influence population distribution in the coming decades (IPCC 2007). Hence, the relationship between climate change, environmental change and human mobility is at the core of numerous studies across diverse fields of research (Warner 2011; McAdam 2010).

Mountain areas cover about 20 % of the Earth's surface, and accommodate approximately 10 % of the global population. Furthermore, they provide key natural resources such as water, energy, minerals, forest and agricultural products to approximately half of the world's population. Additionally, mountains represent essential repositories for the world's ecosystem and' biological diversity, and serve as natural habitats for endangered species (Godde et al. 2000; Smethurst 2000; Viviroli et al. 2007). However, because of their inaccessibility and high dependence on small scale agricultural farming activities in marginal areas, mountains in the southern hemisphere are highly sensitive to the adverse impacts of climate change (Beniston 2003; IPCC 2013, 2014; Jodha 1992; Messerli et al. 2004). Indeed,

---

N. Cascone (✉) • A. Milan
Institute for Environmental & Human Security, United Nations University,
Bonn, Germany
e-mail: noemi-cascone@hotmail.com; milandrea@hotmail.com

A.E. Peña del Valle Isla
Research Programme on Climate Change, National Autonomous University of México,
México, Mexico
e-mail: anepvalle@gmail.com

around 73 % of mountain inhabitants are located in rural areas where they are highly dependent on natural resources for their livelihood. Moreover, their livelihood is subject to physical and economic isolation, land steepness and fragmentation, and low temperatures. However, comparatively few studies have explored the relationship between climate change, environmental change, drivers of livelihood change and human migration in such areas (Kollmair and Banerjee 2011; Milan and Ho 2014; Milan and Ruano 2014; Milan et al. 2015a).

The most comprehensive review on migration and global environmental change to date further points out that in the future, environmental change will significantly impact demographic, economic, political, and social drivers of migration. While the pressure to move is likely to increase, climate change is expect to cause some migration flows to rise but others will decline because people will not be able to leave even if they want to (Foresight 2011).

Recent studies have often focused on migration as part of positive adaptation strategies (Afifi et al. 2015; Bardsley and Hugo 2010; Black et al. 2011; McLeman and Smit 2006; Tacoli 2009). In this chapter, we look at migration as part of successful risk management strategies. In particular, we argue that sending one or more migrants as a risk management strategy at the household level can allow the rest of the household to stay where they are.

This chapter contributes to the literature by exploring the specific case of Las Palomas, a village belonging to the municipality of Xichù in a mountainous region called "Sierra Gorda de Guanajuato", Central Mexico. Sierra Gorda de Guanajuato represents a hotspot of biodiversity (CONABIO 2012) and its ecosystems contain both a rich diversity of species and a high level of redundancy of ecological functions. However, despite its richness in natural and cultural resources, Las Palomas is socially, economically, and environmentally highly vulnerable to climate change. In fact, the majority of its inhabitants are highly dependent on natural resources for their livelihood.

This chapter is mostly based on primary data collected in December 2013 during and after the CATALYST-Local[1] Winter School on "Disaster Risk Reduction and Climate Change Adaptation: Applying science and strategies at the community level". During the Academy, primary data were collected using a grounded theory approach through two transect walks, ten open interviews and group model building. The latter technique refers to a data analysis process in which a client group (in this context, interviewees) involved in the issue at stake is directly involved in the process of model construction. This technique enables the development of new insights, possible new strategies or scenarios (Richardson and Andersen 1995; Richmond 1997; Vennix 1996, 1999; Hare 2011). After the Academy, one of the authors of this chapter conducted seven open interviews with local community members focusing on migration patterns.

---

[1] This project has received funding from the European Union's Seventh Framework Programme for research, technological development and demonstration under grant agreement n°6611188.

The findings of this research demonstrate that migration is mainly perceived by locals as a negative process which leads to the breakdown of families and to the loss of collective memory, in particular regarding local agricultural traditions. Furthermore, the findings of the group work suggested that Las Palomas depends excessively on remittances and social support programs which in turn depend on migration policies (remittances) and governmental spending (social support) which could change at any point in time. Additionally, according to participants of the focus group discussions, this over-reliance on external sources of support has hindered the development of local production capacity and livelihood diversification options.

4.2 Literature Review

4.2.1 Migration as Risk Management Strategy in the Context of Global Environmental Change

In line with the New Economics of Labour Migration (NELM), migration is understood as a risk management strategy adopted mostly at the household level (Stark and Levhari 1982; Stark and Bloom 1985). Moreover, this chapter builds on five key points which emerge from the recent literature on migration in the context of environmental and climate changes (Milan et al. 2015a):

- Environmental change will have an increasing impact on migration in the future through its interrelationship with other demographic, economic, political and social drivers of migration and in the context of rising national inequalities (Foresight 2011). Hence, migration decision-making is always complex and researchers should be careful in establishing any direct relationship between climatic and environmental stressors and migration (Afifi 2011; Bettini 2013; Mortreux and Barnett 2009; Piguet 2012; Wrathall 2012);
- Most migration in the context of environmental change is and will be internal rather than international, with the notable exception of border areas (including mountains) and small states (particularly small island developing states) (Adamo and Izazola 2010; Hugo 1996);
- While migration is often understood and framed as a failure to adapt to environmental and climatic changes, it can also be part of successful livelihood risk management strategies (Afifi et al. 2015; Bardsley and Hugo 2010; Black et al. 2011; McLeman and Smit 2006; Tacoli 2009; Warner and Afifi 2014);
- In the upcoming decades, millions of people who would like to move might be unable to leave locations in which they are vulnerable to environmental change (Black and Collyer 2014; Murphy 2015).
- Existing legal protection gaps should be filled, especially in the case of people displaced across borders in the context of disasters and the effects of climate change (Kälin 2012).

4.2.2 Migration in Mountain Areas of Latin America

Most research on migration in the context of environmental change addresses specific case studies and diverse scientific communities have studied these complex issues through their specific disciplinary lens (McAdam 2010). This has led to the absence of comprehensive theoretical and empirical approaches, which are vital for both research and policy design.

In the face of external stressors, migration, particularly labor migration, represents a frequently used household strategy to enhance and secure varied sources of revenues. Consequently, mountain areas experience important migratory movements, both between rural areas, and to urban and even international destinations. For instance, especially in the off-farm season, farmers in countries such as Peru, Nepal and Pakistan seek seasonal non-farm work in urban areas or abroad (Foresight 2011). Evidence from rural areas of Mexico suggests that climate change especially influences international migration (Nawrotzki et al. 2015a).

Findings from mountainous areas of Latin America show situations where one or more household members migrate while the rest of the households remains. Building on the remittances received, the remaining household members diversify their livelihood in their community of origin. As a consequence, entire households only migrate when local diversification options in the home community run out (Milan and Ho 2014; Milan and Ruano 2014).

A recent study conducted in rural Ecuador also supports that rural out-migration leads to positive outcomes on agricultural activities, causing an *expansion* of agricultural area (Gray and Bilsborrow 2014). However, the authors find conflicting results regarding the impact of remittances. Besides their positive effect, those transfers can have countervailing negative effects as they can substitute agricultural production and effort.

Several studies have explored the relationship between environmental stressors and migration responses in Latin America, as well as the consequent impacts in terms of livelihoods in the communities of origin. A study in rural Ecuador demonstrates that although environmental factors are likely to impact migration flows, the effect is nonlinear. Common narratives suggest that further environmental pressure could trigger large out-migration movements. However, the study reveals a more complex interaction, which in some instances, may even reduce migration flows. Consequently, the authors show that households have a certain agency level and react differently to environmental factors (Gray and Bilsborrow 2013). Data from the southern Ecuadorian Andes suggest that adverse environmental stressors and landlessness do not systematically lead to an increase in out-migration flows. The same study finds that the effects of land ownership and other factors depend on the type of destinations; local, internal, or international (Gray 2009). Additionally, a study on the role of glacial retreat on human migration in the Bolivian Andes finds that glacial retreat has not had an important influence on migratory patterns. However, the research indicates that locals are significantly concerned that glacier disappearance will threaten their livelihoods in the future. Hence, such an event could eventually influence migration decisions in the long-term (Kaenzig 2014). Finally, findings from rural areas in Mexico indicate that social networks could contribute to climate change adaptation in place and hence, may reduce the sensitivity of migration to climate stressors (Nawrotzki et al. 2015b).

4.2.3 Circular Migration as a Risk Management Strategy

Circular migration schemes are gaining growing attention in the policy and research spheres and have led to the development of numerous initiatives in countries such as Canada, Germany and the UK (Foresight 2011).

The Global Commission on International Migration declared that such schemes offer significant development opportunities by allowing migrants to move back and forth more easily (Global Commission on International Migration 2005: 31). The Commission also suggested establishing mechanisms and channels to facilitate migrant's mobility between the country of origin and destination. Further, well managed circular migration programmes can bring "win-win-win"; benefits for the receiving countries (through the reduction of market shortages and through the mitigation of unauthorized migration), for the home countries (through remittances) and for the migrants themselves (through their employment and hence salary) (Vertovec 2008). According to a study on the Canadian Mexican Seasonal Workers Programme, remittances and money accumulated are often invested into the migrants' and their households' land, business, education, housing and health, bringing about positive development to the origin communities (Basok 2003).

Additionally, a growing field of literature provides evidence of circular migration being associated with lower vulnerability levels (McLeman and Hunter 2010; de Moor 2011). Temporary, circular and seasonal migration can represent a valuable income diversification strategy, especially in the face of land degradation and desertification. It could therefore be expected to be increasingly used to cope with slow-onset environmental degradation (de Moor 2011). A recent study conducted by two of the authors of this chapter in La Marmelade, a mountainous community of Haiti subject to long-term degradation, further suggests that individuals engaging in circular migration belong to less vulnerable households than households without such movers (Milan et al. 2015b, forthcoming).

Potential migrants often have recourse to social connections, such as from current or return migrants, to obtain information related to their experience of the destination, or about the work therein. This form of reciprocal assistance can be crucial to the potential migrants and their household. Moreover, by enabling the mobility of migrants, circular migration can contribute to the maintaining of close ties between migrants from afar, their household members, as well as with their community of origin. The technological progresses of the last decades also facilitate virtual transnational connections and mobility (Vertovec 2008). By the same token, this type of migration enables migrants to not abandon their economic activities at home, such as agricultural land and family farming activities, which can be essential to the community of origins' livelihoods (Milan and Ho 2014). Consequently, traditional customs such as agricultural methods can be continued, enhancing the permanence of community-level collective knowledge.

A recent study explores the migrants' role in environmental management in Mexican indigenous communities in Ocumicho, Mexico. In line with previous findings on migrant-sending communities in other regions of Mexico (Jones 2009, 2014; Robson 2010). Lira et al. (2015) suggest that more permanent forms of migration contribute to the deterioration of migrant investments in the community of origin with respect to circular migration. Moreover, the study finds the interrelation between migration and remittances and the consequent impact on agriculture nonlinear, mixed, and complex as the reduction in population caused by out-migration not only lessens the pressure on resources, but it also alters that pressure in a tangled manner. These communities are being transformed by high out-migration flows and changing resource practices. Moreover, the article sheds light on the fact that heightened US border enforcement transforms former circular forms of migration into more permanent ones, which complicates the support that migrants can provide to their communities of origin. Border reinforcement also prevents them from regularly visiting their families back home and returning to the USA. They thus commit to a permanent life there, gradually losing contact with and/or influence on the communities of origin, which can sometimes be reflected in the cease of remittances-sending (Lira et al. 2015). From this conclusion we can derive that investing in legal circular migration schemes could allow for a greater contribution and (financial) investments of migrants from afar to their community of origin. Ultimately, this involvement can then contribute to risk mitigation, adaptation and income diversification strategies in the face of external stressors.

However, while several studies find that the experience and revenues accumulated abroad can provide migrants with better jobs (both in the destination and home areas), others show that circular migrants have the tendency to be trapped in low-skilled types of employment (Vertovec 2008). Authorized migration schemes may attract the same migrants returning year after year to perform the same tasks rather than giving the possibility to migrants to gradually obtain better job opportunities. Additionally, in order for the households to obtain and maintain the lifestyle which the remittances sent by the circular migrants enable, the migrants often continue participating in these migration schemes, fostering a certain dependency and facilitating potential forms of abuse and exploitation from the employer in the destination areas (Ibid).

Furthermore, the continuation of this type of labour migration depends on an array of interrelated factors such as the labour force demand, the economy, the political situation and migration policies in both the destination and the origin areas/countries. Hence, these schemes are volatile in nature and can be disrupted depending, for instance, on economic downturns or on a change in the main political will. Another question that needs to be considered in the elaboration of circular migration schemes is whether they would limit migrants to their current employer or sector, hence contributing to potential exploitation by the employer. Moreover, this form of migration is further embedded in the wider integration debate as these temporary stays preclude integration possibilities and could lead to social exclusion and vulnerability (Ibid). Evidence also suggests that massive male outflows can lead to the worsening of the poverty level and shortage of labour in the sending communities. These risks can however be partially outweighed when incomes are sufficiently high to allow frequent remittances (McLoughlin et al. 2011). Hence, caution needs to be taken in order to establish well-conceived circular migration programs that foster development opportunities for the home and receiving communities as well as reduce the potential negative outcomes connected with such schemes.

4.3 Research Area

Las Palomas lies in the county of Xichú, in the Central Mexican State of Guanajuato. The community is located at 1,282 m above sea level within the Sierra Gorda de Guanajuato ("Fat Mountains"), a dry mountainous region with an average yearly rainfall of 617 mm per year. The number of inhabitants has been slightly decreasing since 2000 when the community counted 319 people, against 309 in 2005 and 277 in 2010, with an average of 4.62 persons per household (Gobierno del Estado de Guanajuato 2013).

Las Palomas was chosen because of its rich ecosystems and its location in an isolated mountainous region of the global South. Other key criteria were its apparent vulnerability to external stressors such as climate and environmental change.

4.3.1 Socio-economic Profile of Las Palomas

The municipality of Xichú (population 11,560) remains affected by poverty and deprivation at significant levels which have prevailed in the region for years (CONANP 2005; INEGI 2011a). In spite of efforts carried out to improve public services and gravel roads, public statistics reveal a historical pattern of underdevelopment associated with an elevated degree of marginalization. According to the National Council of Population, people living in this region have serious deficiencies in their livelihoods, many of them associated with poor basic services, sanitation facilities and infrastructures. Another factor is very low incomes, which can contribute to associated social problems that destabilize households (CONAPO 2011) (Table 4.1).

Opportunities for employment in the region are scarce; they are not well paid and often just seasonal. People working in the agriculture sector represent more than half of the economically active population of the region and they either work as farmers, day labourers or occasional workers. Commercial activities related to public or personal services, construction, industry, transport, as well as the supply of foodstuffs and other provisions also offer employment to the rural labour force (INEGI 2010b).

Table 4.1 Socio-demographic Indicators of Las Palomas

Socio-demographic indicators (2010)	Las Palomas	Xichú (municipality)
Total population	277	11,560
Total households	60	2,655
Average persons per household	4.62	4.4
Women/men	49.8%/50.2%	52.1%/47.9%
Altitude	1,282 m	Between 600 and 2,700 m

Source: INEGI (2009, 2011b) and Gobierno del Estado de Guanajuato (2013)

In environmental terms, the Sierra Gorda lies at the point where the global biogeographical zones of the Neartic and the Neotropical ecozones[2] meet, which means its biodiversity is one of the highest in the world; a place where southern cactus and northern pine live side by side.

In addition, the close-knit communities of the region are part of the broader Huasteca culture of the Mexican central east. Livelihoods have been traditionally built on semi-subsistence agriculture, through a system of Milpas,[3] combined with important income-generating sale of surplus food to local markets.

4.3.2 Climate Threats and Vulnerability in Las Palomas

Despite its richness in natural and cultural resources, Las Palomas is socially, economically, and environmentally, highly vulnerable to climate change risks. In normal years, Las Palomas is characterized by a long "dry" season and another shorter very "rainy" one. The tail-ends of tropical hurricanes also pass through in the hurricane season. Unfortunately, in this semi-arid region, droughts are increasing in frequency and length, reducing water supply for agriculture, and increasing intercommunity conflict over remaining communal spring-water (IEE 2008). This conflict is sometimes reflected by the overnight placing and removing of pipes accessing water from those springs for individual use (fieldwork interviews, December 2013). Additionally, in last decades, Mexico's forests experienced heightened pressure from immoderate and unsustainable use (Lira et al. 2015).

The resulting threats to livelihoods are leading to increased seasonal or permanent migration out of Guanajuato state or abroad, and the breaking up of family

---

[2] The Neartic and the Neotropic ecozones are biogeographical realms. Major habitat types are divided into several biogeographical realms which represent the unique fauna and flora of the world's continents and ocean basins (WWF n.d.). The Nearctic ecozone extends at north of the tropic of Cancer (including the highlands of Mexico), while the Neotropical ecozone extends southwards (*including* Eastern Mexico) (Fund 2014).

[3] A combination of diverse varieties of maize (main crop), along with beans and a variety of squash (known as calabaza), and also chile and quelites (edible leaf vegetables) (Ortiz-Timoteo et al. 2014). Maize is the only crop that follows "a defined spatial order while the other crops are sowed randomly" (Milan and Ruano 2014).

networks (Aguilera and Gómez 2013). This in turn reduces the local memory of agricultural practices and past agreements about water sharing, leading to more conflict. Moreover, reduced superficial and subterranean flows of water are increasing the concentration of naturally occurring arsenic, mercury and lead in the remaining water supply, with as yet unknown impacts on the local communities (CONANP 2005). Further pressure on water supplies is caused by the build-up of organic matter in eutrophying[4] reservoirs, and the diversion of water out of the sub-basin to feed large cities and tourist areas (CEAG 2000).

Moreover, extreme rainfall and hail events are increasing, causing landslides and increased erosion in drought-affected land in typically steep V-shaped valleys, as well as reducing agricultural yield. The increased unpredictability of the weather and the seasons causes uncertainty among farmers as when to sow seeds and harvest, further threatening the sustainability of the communities in this region (Cruz 2013).

Despite the vulnerabilities of Xichú, the Sierra Gorda de Guanajuato region has been largely ignored by development agencies and non-governmental organizations (NGOs) in favour of vulnerable zones in and around Chiapas and Oaxaca, in the south of the country.

4.3.3 Migration in Las Palomas

The State of Guanajuato is one of the states with the greatest number of migrants moving outside of Mexico (Presidencia Municipal de Xichú 2010; INEGI 2010a, b). In 2010, the Census of Population and Housing revealed that Guanajuato was the state with the highest number of migrants at the state level with 119,706 individuals who left Mexico in 2010 (2.2 % of the population of Guanajuato in 2010), representing 10.8 % of all out-migrants from Mexico in that year (INEGI 2010a). Among them, 97.1 % (116,235 people) left their home in Guanajuato in that year to move to the United States.

The National Population Council estimated that approximately 985,000 migrants from Guanajuato are currently living in the United States, and that half of them are there unauthorized. The municipality of Xichú is among the municipalities with highest rates of migration intensity in Guanajuato (Gobierno del Estado de Guanajuato n.d.) and in 2010, occupied ninth position out of 46 municipalities (CONAPO 2012).

Estimates suggest that one in three households in Guanajuato either receives remittances, or they currently have or used to have a family member living in the United States (Gobierno del Estado de Guanajuato n.d.). In 2010, 7.7 % of house-

---

[4] "Eutrophication is a syndrome of ecosystem responses to human activities that fertilize water bodies with nitrogen (N) and phosphorus (P), often leading to changes in animal and plant populations and degradation of water and habitat quality" (Cloern 2013).

holds from Guanajuato received remittances, against 9.2% in 2000 (BBVA 2014). In Xichú however, 26.3% of the households received remittances in 2010 (CONAPO 2012). In the last 5 years, 5.9% of households in Xichú had returning migrants, compared to 4.4% for the whole state of Guanajuato. For the same period of time, another 2% had circular migrants in Xichú against 2.3% for Guanajuato, and 17.3% had migrants to the United States of America as opposed to 5.3% for the whole state of Guanajuato (CONAPO 2012).

Guanajuato is one of the top remittance receiving states. In 2013, it ranked second on the national scale of top receiving states with 2.05 billion dollars in remittances from abroad, behind Michoacán (BBVA 2014). This represents an increase of 18.6% compared to what was recorded in 2004 (Ibid). Whereas, in almost half of the households, remittances represent up to 50% of income, for about a third of the remittance-recipient households such transfers represent total income (Gobierno del Estado de Guanajuato n.d.)

4.3.4 Social Support Schemes

Numerous social initiatives support the inhabitants of rural mountainous communities such as Las Palomas. This section reviews a few of the most established and known programmes.

4.3.4.1 Procampo

Procampo (Program for Direct Assistance in Agriculture) was introduced to compensate for the anticipated negative price effects of trade liberalization on basic crops, to facilitate the transition under The North American Free Trade Agreement (NAFTA) to more liberal policies compared to the previous system of guaranteed prices (Juarez and Hansen 2013). The program provides direct payments per hectare to landowners who registered their land in the Procampo directory (Jones et al. n.d.; Juarez and Hansen 2013). During 2007, about 11.6 million hectares were registered. Originally, the program was planned to end during the 2008–2009 fall-winter crop cycles. Nevertheless, the program was extended and changed in 2014 into Proagro Productivo. Under the new programme, growers may only receive support for planted area, and not just for land ownership (SAGARPA 2014).

4.3.4.2 Progran

The Progran is a tool to support the income of small livestock producers. This was elaborated in 2003 to promote extensive farming practices which would create incentives for increased production of forage in pastures and meadows of the country (factory farming is excluded). This program runs through an animal identification system making Progran payments conditional to the registration of animals to livestock census (Jones et al. n.d.).

4.3.4.3 Prospera

Prospera is the main program of the Mexican government aiming at combating poverty (World Bank 2014; Parker 2003). It is an interagency program that involves the Secretariat of Public Education, the Ministry of Health, the Mexican Social Security Institute, the Ministry of Social Development, and the state and municipal governments (Parker 2003). The program's objective is to help vulnerable families in rural and urban communities, often caught in a vicious circle of intergenerational transmission of poverty to invest in human capital and improve the education, health, and nutrition of their children. The main goal of the program is to focus on the long-term improvement of vulnerable people's economic future.

Prospera is based on the assumption that while poor families are conscious of the potential benefits of investing in their children' education, they often cannot afford to send them to school and would therefore favor taking their children out of school at an early age and send them to work (Parker 2003). Hence, Prospera provides cash transfers to households (subject to certain requirements such as regular school attendance and health clinic visits) to send their children to school instead (World Bank 2014, Ibid).

4.3.4.4 3×1 Migrants Program

The 3×1 Program for Migrants supports the initiatives of Mexicans living abroad and gives them the opportunity to channel resources in the form of remittances to Mexico hence contributing to a certain social impact that directly benefit their communities of origin (IME 2015). The program works in association with clubs or federations of migrants living abroad, and municipal, state and Federal governments – through SEDESOL (The Mexican Social Development Secretariat). Every peso contributed by migrants is matched by a contribution of 3 pesos (one from each government: the Federal, state and municipal governments). The program created strong links between communities of origin and communities of destination by allowing migrant clubs to become new players in the promotion of local and regional development and policy design through their role as interlocutors with the three levels of government. The main contribution of 3×1 is social learning based on an alternative conception of development driven by social change rather than by the logic of the market (García Zamora 2005). The results of the program have been varied and its success is often linked to the level of trust between migrant organizations and the governments. In 2013, 160 projects and actions were performed in 28 municipalities of Guanajuato, amongst which 28 were considered "productive" projects that helped women in communities develop micro and small enterprises (Sánchez 2014).

4.3.4.5 Seasonal Agricultural Worker Program (SAWP)

Representing 11% of all migrant workers, Mexico was the second largest pool of temporary workers in Canada in 2011. Most Mexican temporary workers (94%) work there under the Seasonal Agricultural Worker Program (SAWP). Since 1974, Mexican workers under this program are allowed to be employed for a maximum of 8 months on Canadian farms (Massey and Brown 2011).

Recruited through a government-to-government memorandum of understanding (MoU), most SAWP migrants (around 80%) work on fruit, vegetable and tobacco farms and stay on average for 4 months (Massey and Brown 2011).

The program is administered by an office located in the capital and applicants have to visit the office numerous times during their application. As a consequence, participants originate mainly from rural regions near Mexico City. In 2005, for instance, about a quarter of laborers originated from the state of México, compared to only 7% from Guanajuato (Massey and Brown 2011). The number of participants has increased progressively, with the majority of laborers engaged in circular migration.

4.4 Adaptation Processes in Las Palomas

4.4.1 A Resilient Human-Environmental System Based on Ecosystem Services

The socio-cultural system supporting communities like Las Palomas is very complex and in a process of change. Approaching adaptation in Las Palomas requires looking at far more than adjustments in infrastructure and cultivation. Adaptation has to be the result of a complex mix of a resilient environment, adaptive capacities and the sustainability of present-day adaptive strategies.

Ecosystem services are the multiple types of benefits that robust ecosystems provide to humankind. An ecosystem that supports redundancy of ecological functions produces robust ecosystems (Peña del Valle 2014) and, by extension, robust human ones. Having built-in redundancy in the ecosystem (i.e. an ecosystem service such as food provision can be provided by more than one ecosystem element), means that humans can build-in redundancy in their livelihoods in the form of diversification of productive activities. Multiple options mean that, in case one element or function is lost, another one can replace it (Peña del Valle 2014). This enables human-environmental systems to accommodate change and to ensure their survival (Aguilera and Gómez 2013).

The region of Sierra Gorda has ecosystems that contain both a rich diversity of species, and a high level of redundancy of ecological functions, representing a hotspot of biodiversity (CONABIO 2012). For Las Palomas, this is the foundation of a resilient human-environmental system which can buffer the community from external disturbances (e.g. droughts, creeping soil erosion, pests, loss of incomes in a failing market economy, increased market competition) (Cruz 2013) and if necessary, allow the community to recover from potential losses in livelihoods.

More specifically, many of the ecosystem services available to Las Palomas, such as the conservation of fertile soils and water purification provided by the surrounding forested areas, help save money and time spent on local productive activities. By keeping forested areas in between croplands, farmers can use those areas to prevent soil erosion and compaction, which because of the maintenance of soil fertility, leads to a reduction in expenses for fertilisers and manual labour (fieldwork interviews, December 2013). Moreover, these forested areas keep populations of birds and insects high, which act as pollinators and seed dispersers (Ramírez-Albores et al. 2015).

Diverse ecosystem-based livelihoods help to ensure food and income security. For example, in addition to the *Milpa*, the traditional crop-growing system for sustaining the family, women in Las Palomas can diversify their subsistence production into crops with a market value, by growing nuts and agave, as well as various spices (e.g. oregano) and medicinal plants. They can also gather wild mushrooms, pine nuts, and firewood (fieldwork interviews, December 2013).

4.4.2 Adaptive Capacities: Social Engagement and Local Institutions

4.4.2.1 Dialogues for Resource Management

Uncertainty makes planning and decision-making more difficult for people in Las Palomas and, traditional adaptive strategies may be no longer sufficient. Thus, people have developed the capacity to improvise as they go while they strengthen their capabilities for responding and adjusting to disturbance and risk. This enables people to keep their livelihoods dynamic and open to change in a way that has relevance to their needs. For example, inhabitants of Las Palomas have proved to be able to reach agreements with the neighbouring community "El Toro" on water usage and distribution between the two communities. In May 2012, during the middle of the dry season, an issue about water rights came about and the two communities engaged in dialogue and reached an agreement (fieldwork interviews, December 2013).

4.4.2.2 Promotion of Flexibility and Diversification

Another example is the ability to bring new ideas that promote livelihood diversification and savings. The local health centre has been suggesting households to retake their backyard orchards and to grow several vegetables that would be central to improve people's diet. In this respect, such initiatives consider that people would be able to diversify their income base as well as to save by using domestic resources cooking, healing, and building their houses. Also, in case of disaster, when roads are usually damaged and transportation of goods and people is difficult, people could ensure subsistence production and improved food security (fieldwork interviews, December 2013).

4.4.2.3 Low-Tech Early Warning Systems

Low-cost, low-tech early warning systems are another example of actions which are needed, although local capacity is insufficient in terms of social awareness and collective engagement to facilitate it. In terms of infrastructure, Las Palomas has no mobile phone connection and extremely limited (or no) access to the internet. Moreover, lack of funds and the remoteness of Las Palomas also imply that the community needs low tech solutions for heavy metal remediation methods, and erosion control (fieldwork interviews, December 2013).

4.4.2.4 Ability to Create Low-Tech Solutions

People in Las Palomas would benefit from innovative technological approaches to increase the quantity and quality of water in the region for domestic and agricultural use. The high altitude of the area means that dew capture is a possibility, and is practiced successfully by some people who are able to capture an average of 48 l per night (fieldwork interviews, December 2013).

4.5 Migration and Adaptation: Results of Fieldwork

4.5.1 Migration as Part of Adaptation Strategies

In Las Palomas, around 10–15 individuals (out of 277 inhabitants) leave the community every year, usually for 6 months (between December and June) under seasonal migration programs (fieldwork interviews, December 2013). Those programs usually attract the same individuals every year, meaning the community is not depopulating. Those programs are becoming more and more established in the community, attracting a limited yet increasing number of individuals. Another mobility pattern observed in the region regards short-term migration for a period of a few weeks to the city of Queretaro for construction work. Only few people move to Mexico City, unlike in the 1970s and 1980s when most migrants of the region left for the capital (fieldwork interviews, December 2013).

Data from the semi-structured interviews on migration suggests that seasonal migration programs from Mexico to the United States have provided households with increased income through remittances sent by their household member(s) abroad. This financial help further enabled household to better plan their expenses. Additionally, seasonal schemes to the United States have contributed to the reduction of irregular migration-related risks (fieldwork interviews, December 2013).

However, the process of migration is perceived as leading to some negative consequences such as family breakdown and loss of social values, but also loss of the collective memory of traditional and resilient methods of agriculture. Some of the migrants do not return to their communities of origin but settle abroad, and gradually lose ties with their community of origin. Hence, in this case, their families, in addition to experiencing the loss of a loved one, do not receive remittances anymore as compensation mechanisms (fieldwork interviews, December 2013).

Migration can represent yet another process by which locals adapt to external stressors and minimize livelihood risks. In the case of circular migration, migrants tend to be away from their families for a limited amount of time only. The maintenance of family ties does by the same token foster the continuity of the sending of remittances. Additionally, those migrants often return home with social remittances, such as the skills learned on the spot, which can then benefit the entire community. Moreover, those circular schemes enable families to remain and adapt in-situ. As a result, this process ensures the maintaining and fostering of collective memories, which can take the form of traditional agricultural practices.

4.5.2 Barriers to Local Adaptation

Based on fieldwork observations, the inhabitants from Las Palomas possess a wide and complex range of adaptive capacities, which are developed, or ignored, to different degrees according to the individual perception, experiences and decisions of those inhabitants. While some of the existing capacities allow people to benefit from a robust ecosystem that provides a variety of services to people in Las Palomas, other capacities available in Las Palomas are not used. In fact, local people often do not follow successful examples of local adaptation employed by other members of the same community.

Barriers to local adaptation relate to the way people and households perceive risks and opportunities, and how they integrate them into their livelihood decision-making process (Litre and Burzstin 2015). Some of these barriers may also relate to a lack of knowledge, economic factors, social or cultural conflicts, or simply to an aversion to change (Moser and Ekstrom 2010).

For example, advice from the health centre about adapting crop choice and diet are ignored due to a lack of knowledge about nutrition, or due to strong cultural connections to the traditional milpa system that override any perceived health risk from continuing a traditional diet. Crop diversification might also be prevented by the increased level of climate change-induced uncertainty related to cropping and harvesting schedules. The perceived likelihood of unexpected events such as a hail storm destroying a crop before harvesting, could disincentivize farmers to take the risk of adapting their crop.

Finally, livelihood diversification options can also become over time a barrier to adaptation in itself. The over-reliance of Las Palomas on external sources of support, such as remittances and governmental support, has limited their local production capacity as well as the development of other local livelihoods diversification options. Remittances may be perceived, rightly or wrongly, to be a stable, long term substitute for local production and diversification. However, labour migration depends on policy decisions and economic considerations which are exogenous to the community. As a consequence, it is important that household livelihoods do not solely rely on external sources of revenues such as remittances sent by migrants.

While this chapter discusses the potential of circular migration as a risk management strategy, heavy reliance on such schemes could have negative effects. However, this chapter highlights that well-conceived circular migration schemes, combined with the reinforcement of traditional agricultural methods could build community resilience and allow the community to pursue a climate-resilient development pathway.

4.6 Conclusion

This chapter investigated the role of circular migration within livelihood diversification and risk reduction strategies at the household level in Las Palomas. Moreover, it attempted to fill a knowledge gap on the potential of circular migration as well as on the limits and barriers to adaptation in isolated mountain areas in the context of climate change and environmental change.

Based on primary qualitative data, the study highlighted the potential of well-conceived circular migration schemes for the improvement of the resilience of the Mexican community of Las Palomas to external stressors. The findings suggest that sending one or more migrants abroad as a risk management strategy at the household level can enable the rest of the household to stay where they are, increase their resilience through increased income and reduce livelihood risks. Additionally, this chapter stresses the importance for households to access both options for safe and orderly migration as well as local agricultural diversification options, without excessive reliance on either.

References

Adamo, S. B., & Izazola, H. (2010). Human migration and the environment. *Population and Environment, 32*(2–3), 105–108.

Afifi, T. (2011). Economic or environmental migration? The push factors in Niger. *International Migration, 49*, e95–e124. doi:10.1111/j.1468-2435.2010.00644.x.

Afifi, T., Milan, A., Etzold, B., et al. (2015). Human mobility in response to rainfall variability: Opportunities for migration as a successful adaptation strategy in eight case studies. *Migration and Development*. doi:10.1080/21632324.2015.1022974. http://www.tandfonline.com/eprint/FStEr75WNnyCs9my4Cfp/full. Accessed 24 Feb 2016.

Aguilera, A., & Gómez, C. (2013). Economía, Población, y Migración en la Sierra Gorda. Cuaderno de Trabajo. Universidad de Guanajuato. 27 p.

Banco Bilbao Vizcaya Argentaria, S. A. (BBVA). (2014). Migration outlook: Mexico. First half 2014 economic analysis 2014, joint project of BBVA Bancomer Foundation and Mexico Economic Research Department of BBVA Research. Translation from: Situación Migración México (2014). Primer Semestre 2014 Análisis Económico ISSN 2007-6932. https://www.bbvaresearch.com/wp-content/uploads/2014/08/1407_SitMigracionMexico_1S14.pdf. Accessed 24 Feb 2016.

Bardsley, D. K., & Hugo, G. J. (2010). Migration and climate change: Examining thresholds of change to guide effective adaptation decision-making. *Population and Environment, 32*(2–3), 238–262. doi:10.1007/s11111-010-0126-9.

Basok, T. (2003). Mexican seasonal migration to Canada and development: A community-based comparison. *International Migration, 41*(2), 3–26.

Beniston, M. (2003). Climatic change in mountain regions: A review of possible impacts. *Climatic Change, 59*, 5–31.

Bettini, G. (2013). Climate barbarians at the gate? A critique of apocalyptic narratives on 'climate refugees'. *Geoforum, 45*, 63–72.

Black, R., & Collyer, M. (2014). Populations 'trapped' at times of crisis. *Forced Migration Review, 45*, 52–56.

Black, R., Bennett, S. R. G., Thomas, S. M., & Beddington, J. R. (2011). Climate change: Migration as adaptation. *Nature, 478*(7370), 447–449.

Cloern, J. (2013). Eutrophication. The encyclopedia of Earth. http://www.eoearth.org/view/article/51cbedc37896bb431f693ba8. Accessed 24 Feb 2016.

Comisión Estatal del Agua de Guanajuato (CEAG). (2000). Sinopsis del Estudio de Prospección Hidrogeológica del Acuífero de Xichú, Gto. 69 páginas. Archivos en línea de la Comisión EStatal del Agua de Guanajuato.

Comision Nacional de Áreas Naturales Protegidas (CONANP). (2005). Estudio Previo Justificativo para el establecimiento del Área Natural Protegida Reserva de la Biósfera Sierra Gorda de Guanajuato. 281 p.

Comisión Nacional para el Conocimiento y Uso de la Biodiversidad (CONABIO). (2012). estrategia para la Conservación y Uso Sustentable de la Biodiversidad del Estado de Guanajuato. Tomo I Diagnóstico. 72 p.

Consejo Nacional de Población (CONAPO). (2011). Indice de Marginación por entidad federativa y municipio 2010. http://www.conapo.gob.mx/es/CONAPO/Indices_de_Marginacion_2010_por_entidad_federativa_y_municipio. Accessed 24 Feb 2016.

Consejo Nacional de Población (CONAPO). (2012). Anexo B. Índices de intensidad migratoria México-Estados Unidos por entidad federativa y municipio 2010. El estado de la migración. Índices de intensidad migratoria. Colección: índices sociodemográficos. http://www.conapo.gob.mx/work/models/CONAPO/intensidad_migratoria/anexos/Anexo_B1.pdf. Accessed 24 Feb 2016.

Cruz, D. (2013). Adaptación al Cambio Climático en el Área Protegida Sierra Gorda. Tésis de Licenciatura. Facultad de Filosofía y Letras. Universidad Nacional Autónoma de México, 141 p.

de Moor, N. (2011). *Labour migration for vulnerable communities: A strategy to adapt to a changing environment*. Bielefeld: COMCAD. General Editor: Thomas Faist, Working Papers – Center on Migration, Citizenship and Development; 101.

Foresight. (2011). *Migration and global environmental change. Final project report*. London: The Government Office for Science.

Fund, W. (2014) Ecoregion. The encyclopedia of Earth. Accessed 24 Feb 2016.

García Zamora, R. (2005). Collective remittances and the 3x1 as a transnational social learning process. Background paper presented at the seminar "Mexican Migrant Social and Civic

Participation in the United States. Woodrow Wilson International Center For Scholars, Washington DC, November 4th and 5th.

Global Commission on International Migration. (2005). *Migration in an interconnected world: New directions for action*. Geneva: GCIM.

Gobierno del Estado de Guanajuato (GTO). (2013). El Padrón de Comunidades Indígenas del Estado de Guanajuato, monografía Indigenas: Palomas, Xichù. Secretaría de Desarrollo Social y Humano "Guanajuato orgullo y compromiso de todos" (GTO), Impulso Guanajuato. http://portalsocial.guanajuato.gob.mx/sites/default/files/documentos/2013_SEDESHU_xichu-palomas.pdf. Accessed Feb 2016.

Gobierno del Estado de Guanajuato (GTO). (n.d.). Programa Especial De Migración 2013–2018. https://transparencia.guanajuato.gob.mx/biblioteca_digital/docart10/201501131046140. ProgramaEspecialMigracion.pdf. Accessed 24 Feb 2016.

Godde, P. M., Price, M. F., & Zimmermann, F. M. (Eds.). (2000). *Tourism and development in mountain regions*. New York: CABI Publishing.

Gray, C. L. (2009). Environment, land, and rural out-migration in the southern Ecuadorian Andes. *World Development, 37*(2), 457–468. doi:10.1016/j.worlddev.2008.05.004.

Gray, C. L., & Bilsborrow, R. E. (2013). Environmental influences on human migration in rural Ecuador. *Demography, 50*, 1217–1241. doi:10.1007/s13524-012-0192-y.

Gray, C. L., & Bilsborrow, R. E. (2014). Consequences of out-migration for land use in rural Ecuador. *Land Use Policy, 36*, 182–191. doi:10.1016/j.landusepol.2013.07.006. http://DOI.org/10.1016/j.landusepol.2013.07.006. Accessed 24 Feb 2016.

Hare, M. P. (2011). Forms of participatory modelling and its potential for widespread adoption in the water sector. *Environmental Policy and Governance, 21*(4), 386–402.

Hugo, G. (1996). Environmental concerns and international migration. *The International Migration Review, 30*(1), 105–131. Accessed 24 Feb 2016.

Instituto de Ecología del Estado (IEE). (2008). Hacia una Estrategia Estatal de Cambio Climático en Guanajuato. Gobierno del Estado de Guanjuato, 81 p.

Instituto de los Mexicanos en el Exterior (IME). (2015). Progama 3X1. SEDESOL. http://www.ime.gob.mx/programa-3x1. Accessed 24 Feb 2016.

Instituto Nacional de Estadística y Geografía (INEGI). (2009). Prontuario de información geográfica municipal de los Estados Unidos Mexicanos Xichú, Guanajuato, Clave geoestadística 11045.

Instituto Nacional de Estadística y Geografía (INEGI). (2010a). Censo de Población y Vivienda 2010. Cuestionario ampliado. http://www3.inegi.org.mx/sistemas/temas/default.aspx?s=est&c=17484. Accessed 24 Feb 2016.

Instituto Nacional de Estadística y Geografía (INEGI). (2010b). Principales resultados del Censo de Población y Vivienda 2010. http://www.planetaj.cruzrojamexicana.org.mx/pagnacional/secciones/Juventud/Contenido/PlanetaJ/dowloadfiles/CENSO2010_principales_resultados.pdf. Accessed 24 Feb 2016.

Instituto Nacional de Estadística y Geografía (INEGI). (2011a). *Principales resultados del Censo de Población y Vivienda 2010: Guanajuato*. México: INEGI. 84 p.

Instituto Nacional de Estadística y Geografía (INEGI). (2011b). *Censo de Población y Vivienda (2010)*. México: Panorama sociodemográfico de Guanajuato. ISBN 978-607-494-208-8. http://seieg.iplaneg.net/pmd/doc/xichu/i.insumos/3.estadisticas_e_indicadores/xichu.pdf. Accessed 24 Feb 2016.

Intergovernmental Panel on Climate Change (IPCC). (2007). *Intergovernmental panel on climate change: Climate change: Impacts, adaptation and vulnerability*. Cambridge, UK: Cambridge University Press.

Intergovernmental Panel on Climate Change (IPCC). (2013). Intergovernmental panel on climate change: Climate change 2013: The physical science basis. In T. F. Stocker, D. Qin, G.-K. Plattner, M. Tignor, S. K. Allen, J. Boschung, A. Nauels, Y. Xia, V. Bex, & P. M. Midgley (Eds.), *Contribution of working group I to the fifth assessment report of the intergovernmental panel on climate change*. Cambridge, UK: Cambridge University Press. 1535 pp.

Intergovernmental Panel on Climate Change (IPCC). (2014). Intergovernmental panel on climate change: Climate change 2014: Impacts, adaptation, and vulnerability, part a: Global and sectoral aspects. In C. B. Field, V. R. Barros, D. J. Dokken, K. J. Mach, M. D. Mastrandrea, T. E. Bilir, M. Chatterjee, K. L. Ebi, Y. O. Estrada, R. C. Genova, B. Girma, E. S. Kissel, A. N. Levy, S. MacCracken, P. R. Mastrandrea, & L. L. White (Eds.), *Contribution of working group II to the fifth assessment report of the intergovernmental panel on climate change*. Cambridge, UK: Cambridge University Press. 1132 pp.

Jodha, N. S. (1992). Mountain perspective and sustainability: A framework for development strategies. In N. S. Jodha, M. Banskota, & T. Partap (Eds.), *Sustainable mountain agriculture volume 1: Perspective and issues*. New Delhi: Oxford and IBH Publishing Co. http://dx.DOI.org/10.3362/9781780443546.

Jones, R. C. (2009). Migration permanence and village decline in Zacatecas: When you can't go home again. *The Professional Geographer, 61*(3), 382–399.

Jones, R. C. (2014). The decline of international migration as an economic force in rural areas: A Mexican case study. *International Migration Review, 48*(3), 728–761.

Jones, J. J. M., Ochoa, R. F. O. Cabello, P. S. Knutson, R. D., Fernandez, C. C., Knutson, R. D., Westhoff, P. C, & Brown, D. S. (n.d.). Proyecciones para el Sector Agropecuario de México, Escenario Base del Sector Agropecuario en México, Proyecciones 2009–2018, Secretaría de Agricultura, Ganadería, Desarrollo Rural, Pesca y Alimentación Subsecretaría de Fomento a los Agronegocios, Agriculture and Food Policy Center Texas A&M, Food and Agricultural Policy Research Institute University of Missouri, Gobierno federal, SAGARPA. http://www.sagarpa.gob.mx/agronegocios/Documents/Escenariobase09.pdf. Accessed 24 Feb 2016.

Juarez, B., & Hansen, E. W. (2013). PROCAMPO 2013 subsidy program changes USDA Foreign Agricultural Service, Global Agricultural Information Network. http://gain.fas.usda.gov/Recent%20GAIN%20Publications/PROCAMPO%202013%20Subsidy%20Program%20Changes_Mexico_Mexico_2-14-2013.pdf. Accessed 24 Feb 2016.

Kaenzig, R. (2014). Can glacial retreat lead to migration? A critical discussion of the impact of glacier shrinkage upon population mobility in the Bolivian Andes. *Population and Environment, 36*, 480–496.

Kälin, W. (2012). From the Nansen principles to the Nansen initiative. *Forced Migration Review, 41*, 48–49.

Kollmair, M., & Banerjee, S. (2011). *Drivers of migration in mountainous regions of the developing world: A review* (Foresight: Migration and global environmental change driver review 9). London: Government Office for Science.

Lira, M. G., Robson, J. P., & Klooster, D. J. (2015). Can indigenous transborder migrants affect environmental governance in their communities of origin? Evidence from Mexico. *Population and Environment*, 1–15. doi:10.1007/s11111-015-0247-2

Litre, G., & Bursztyn, M. (2015). Perceções e adaptação aos riscos climáticos e socioeconómicos na pecuária familiar do bioma Pamapa. *Ambiente & Sociedade, 18*(3), 55–80. 09/2015.

Massey, D., & Brown, A. E. (2011). Movement between Mexico and Canada: Analysis of a new migration stream. *International Migration, 6*(1), 119–144 (Geneva, Switzerland).

McAdam, J. (Ed.). (2010). *Climate change and displacement: Multidisciplinary perspectives*. Oxford: Bloomsbury Publishing.

McLeman, R. A., & Hunter, L. M. (2010). Migration in the context of vulnerability and adaptation to climate change: Insights from analogues. Wiley interdisciplinary reviews. *Climate Change, 1*(3), 450–461. http://DOI.org/10.1002/wcc.51. Accessed 24 Feb 2016.

McLeman, R. A., & Smit, B. (2006). Migration as an adaptation to climate change. *Climatic Change, 76*, 31–53.

McLoughlin, S., Münz, R., Bünte, R., Hultin, G., Müller, W., & Skeldon, R. (2011). *Temporary and circular migration: Opportunities and challenges*. European Policy Centre, Working Paper, (35).

Messerli, B., Viviroli, D., & Weingartner, R. (2004). Mountains of the world: Vulnerable water towers for the 21 century. *Ambio, 13*, 29–34.

Milan, A., & Ho, R. (2014). Livelihood and migration patterns at different altitudes in the Central Highlands of Peru. *Climate and Development, 6*(1), 69–76. doi:10.1080/17565529.2013.8261 27.

Milan, A., & Ruano, S. (2014). Rainfall variability, food insecurity and migration in Cabricán, Guatemala. *Climate and Development, 6*(1), 61–68. doi:10.1080/17565529.2013.857589.

Milan, A., Gioli, G., & Afifi, T. (2015a). Migration and global environmental change: Methodological lessons from mountain areas of the global south. *Earth System Dynamics, 6*(1), 375.

Milan, A., Melde, S., Cascone, N., Schindler, M., & Warner, K. (2015b). When do household benefit from migration? Insights from vulnerable environments in Haiti: Migration, Environment, and Climate Change: Policy Brief Series: Issue 8, vol (1). http://environmentalmigration.iom.int/policy-brief-series-issue-8-when-do-households-benefit-migration. Accessed 24 Feb 2016.

Milan, A., Melde, S., Cascone, N. (Forthcoming). MECLEP Haïti: Une Analyse de la relation entre la vulnérabilité et la migration dans le contexte d'adaptation à la dégradation environnementale et catastrophes. Rapport: l'enquête auprès des ménages. *International Organization for Migration*.

Mortreux, C., & Barnett, J. (2009). Climate change, migration and adaptation in Funafuti, Tuvalu. *Global Environmental Change, 19*, 105–112.

Moser, S. C., & Ekstrom, J. A. (2010). A framework to diagnose barriers to climate change. *PNAS, 107*, 22026–22031. Early edition.

Murphy, D. W. (2015). Theorizing climate change, (im) mobility and socio-ecological systems resilience in low-elevation coastal zones. *Climate and Development, 7*(4), 380–397.

Nawrotzki, R. J., Hunter, L. M., Runfola, D. M., & Riosmena, F. (2015a). Climate change as a migration driver from rural and urban Mexico. *Environmental Research Letters, 10*(11), 114023.

Nawrotzki, R. J., Riosmena, F., Hunter, L. M., & Runfola, D. M. (2015b). Amplification or suppression: Social networks and the climate change—Migration association in rural Mexico. *Global Environmental Change, 35*, 463–474.

Ortiz-Timoteo, J., Sánchez-Sánchez, O., & Ramos-Prado, J. M. (2014). Productive activities and management of the Milpa in three communities of the Municipality of Jesús Carranza, VERACRUZ, MEXICO. *Polibotánica, 38*, 173–191.

Parker, S. W. (2003). Shanghai poverty conference – Scaling up poverty reduction case study. The oportunidades program in Mexico. http://info.worldbank.org/etools/docs/reducingpoverty/case/119/fullcase/Mexico%20Oportunidades.pdf. Accessed 24 Feb 2016.

Peña del Valle, A. E. (2014). Al mal tiempo, buena resiliencia, *Revista Ciencias*. Enero 2014 (pp. 111–112). Facultad de Ciencias, Universidad Nacional Autónoma de México.

Piguet, E. (2012). Migration: The drivers of human migration. *Nature Climate Change, 2*, 400–401.

Presidencia Municipal de Xichú. (2010). Acuerdo Municipal Mediante el cual se Aprueba el Plan de Gobierno Municipal 2009–2012 del Municipio de Xichú, gto., *Periódico Oficial del Gobierno del Estado de Guanajuato*.

Ramírez-Albores, J. E., Gordillo-Martínez, A., & Navarro-Sigüenza, A. G. (2015). Registros notables y listado avifaunístico en un área de la Reserva de la Biosfera Sierra Gorda de Guanajuato, México. *Revista Mexicana de Biodiversidad, 86*(4), 1058–1064. Universidad Nacional Autónoma de México.

Richardson, G. P., & Andersen, D. F. (1995). Teamwork in group model building. *System Dynamics Review, 11*(2), 113–137.

Richmond, B. (1997). The strategic forum: Aligning objective, strategy, and process. *System Dynamics Review, 13*(2), 131–148.

Robson, J. P. (2010). The impact of rural to urban migration on forest commons in Oaxaca, Mexico. Dissertation, University of Manitoba, Winnipeg, Canada.

SAGARPA. (2014). Transformación de PROCAMPO a PROAGRO acentúa impulso a producción agroalimentaria Mexico D.F. http://www.sagarpa.gob.mx/saladeprensa/2012/Paginas/2014B022.aspx#. Accessed 24 Feb 2016.

Sánchez, J. (2014, February 24). Celaya, Aportará el estado 55 mdp para el programa Migrante: Vargas, La Prensa, Organización Editorial Mexicana S.A. de C.V. http://www.oem.com.mx/laprensa/notas/n3302170.htm. Accessed 24 Feb 2016.

Smethurst, D. (2000). Mountain geography. *Geographical Review, 90*, 35–56.

Stark, O., & Bloom, D. E. (1985). The new economics of labour migration. *The American Economic Review, 75*, 173–178.

Stark, O., & Levhari, D. (1982). On migration and risk in LDC. *Economics Development and Culture Change, 31*, 191–196.

Tacoli, C. (2009). Crisis or adaptation? Migration and climate change in a context of high mobility. *Environment and Urbanization, 21*, 513–525.

Vennix, J. (1996). *Group model building*. New York: Wiley.

Vennix, J. (1999). Group model-building: Tackling messy problems. *System Dynamics Review, 15*(4), 379–401.

Vertovec, S. (2008). Circular migration: The way forward in global policy? *Canadian Diversity, 6*(3), 36–40.

Viviroli, D., Dürr, H. H., Messerli, B., Meybeck, M., & Weingartner, R. (2007). Mountains of the world, water towers for humanity: Typology, mapping, and global significance. *Water Resources Research, 43*, W07447. doi:10.1029/2006WR005653.

Warner, K. (2011). Interdisciplinary approaches to researching environmental change and migration: Methodological considerations and field experiences from the EACH-FOR project. In C. Vargas-Silva (Ed.), *Handbook of research methods in migration* (pp. 366–395). Oxford: Edward Elgar Publishing.

Warner, K., & Afifi, T. (2014). Where the rain falls: Evidence from 8 countries on how vulnerable households use migration to manage the risk of rainfall variability and food insecurity. *Climate and Development, 6*, 1–17.

World Bank (2014). A model from Mexico for the world. http://www.worldbank.org/en/news/feature/2014/11/19/un-modelo-de-mexico-para-el-mundo. Accessed 24 Feb 2016.

World Wildlife Fund (WWF). (n.d). *Ecoregions*. http://www.worldwildlife.org/biomes. Accessed 24 Feb 2016.

Wrathall, D. J. (2012). Migration amidst social-ecological regime shift: The search for stability in Garifuna villages of northern Honduras. *Human Ecology, 40*, 583–596.

Part II
Low-Lying Areas

Chapter 5
Household Adaptation Strategies to Climate Extremes Impacts and Population Dynamics: Case Study from the Czech Republic

Robert Stojanov, Barbora Duží, Ilan Kelman, Daniel Němec, and David Procházka

5.1 Introduction

Climate change impacts are projected to become a serious environmental challenge, and risk management is attracting the attention of international communities and individual states (IPCC 2012; EEA 2012). This leads to actions of not only climate change mitigation strategies (reducing greenhouse gas emissions and increasing their sinks), but also implementing preventive and adaptation strategies for climate change impacts (Wamsler and Brink 2014). Innovative potential also resides in the search for new methods of adaptation management, such as "adaptive co-management for climate change adaptation" (Plummer and Baird 2013), the search

---

R. Stojanov (✉) • D. Procházka
Faculty of Business and Economics, Department of Informatics, Mendel University in Brno, Zemědelská 1, 613 00 Brno, Czech Republic
e-mail: stojanov@centrum.cz; david.prochazka@mendelu.cz

B. Duží
Department of Environmental Geography, Institute of Geonics of the CAS, Drobného 28, 602 00 Brno, Czech Republic
e-mail: arobrab@centrum.cz

I. Kelman
University College London and University of Agder, London and Kristiansand, UK and Norway
e-mail: ilan_kelman@hotmail.com

D. Němec
Faculty of Economics and Administration, Department of Economics, Masaryk University, Lipová 41a, 602 00 Brno, Czech Republic
e-mail: nemecd@econ.muni.cz

for pathways of adaptation measures whilst taking into account the high level of uncertainty (Walker et al. 2013) or bottom-up approaches known as "community-based adaptation" (Reid et al. 2009).

Population dynamics as seen through migration have always been part of human endeavors throughout history. Environmental degradation, resource depletion, natural hazards and natural hazard drivers including climate change play a contributing role as important 'push' factors affecting population movement. The truism is that people do not often move for a single reason. The motivation to move involves a complex web of multiple factors that denote individual belief, pursuit, and dreams, or collective decisions for family or cultural groups, within specific local, national or international economic, social, and political contexts. Human migration can also be seen as a strategy for dealing with a specific situation in life such as job loss, poor harvests, or as a long-term adaptation strategy to a significant change in the environment (such as extreme drought) or loss of housing. However, an examination of the environmental (including climatic) factors that may affect or cause migration has not always been given sufficient consideration in migration discourse and is thus a relatively new research topic.

Population movement in the projected climate change impacts context is frequently understood as a household adaptation strategy for or in response to climate-related extremes (see Tacoli 2009; McLeman and Hunter 2010; Bardsley and Hugo 2010; Black et al. 2011; Oliver-Smith 2012; Warner and Afifi 2014; Stojanov et al. 2014a). The key question is whether migration may enhance the capacity of communities to adapt to climate change, especially as a response to extreme weather. Neumann and Hilderink (2015) point out that climate variability could play some role among other push factors influencing human migration. Kelman et al. (2015) remind us that vulnerability and resilience to environmental hazards including climate change, and manifestations of climate change such as through extreme weather, are indeed reasons for mobility, as well as non-mobility.

Among other natural hazards and their impacts, such as drought, heat waves, earthquakes, hurricanes and others, floods are among one of the most frequent and most serious climate extremes affecting population, infrastructure and settlements. Jongman et al. (2014) show that extreme discharges are strongly correlated across European river basins and present probabilistic trends in continental flood risk. They demonstrate that observed extreme flood losses could more than double in frequency by 2050 under climate change projections and socio-economic development while demonstrating that risk can be shared by expanding risk transfer financing, reduced by investing in flood risk reduction or absorbed by enhanced solidarity between countries.

While floods have always had their place in nature, the expectation under climate change is that, in many locations, they will become worse and more frequent. Building on river flood risk work in the context of complicated river flood trends in Central Europe under climate change, we focus on societal adaptations to river flood risk in Central Europe at the household level. People's perceptions of river flood risk and adaptation responses to that perceived risk can be complex. Residents may

Table 5.1 The main floods in the Czech Republic since 1997

Year	Season	Number of waves	Intensity in general and serious impact on targeted region
1997	Summer floods	2	Q 100- Q 800 (yes)
2002	Summer flood	1	Q 500–1,000 (yes)
2006	Spring flood	1	Q 50–100 (yes)
2009	Summer floods	Several	Around Q100 (yes)
2010	Summer floods	Three	Around Q100 (yes)
2013	Summer floods	Three	Around Q100 (no)

Source: according to data from CHMI (1997, 2002, 2006, 2009, 2010, 2013) (CHMI 2013)
Series of floods, which occurred on several regions in the Czech Republic

choose several adaptation options, depending on their specific situation and conditions. Movement from affected areas is just one of many possibilities within population dynamics.

The Czech Republic is of particular interest in the European context due to the several, recent flooding disasters which were national emergencies, including in 1997, 2002, 2006, 2009, 2010, and 2013 (Duží et al. 2015a; and for more details see Table 5.1), and given that few previous studies seek to understand what Czech people think about flooding and adapting to it. One exception is Duží et al. (2015b) based on 300 households living in 10 municipalities from the upper reaches of Bečva River basin.

Table 5.1 shows the timing, characteristics, and magnitudes of floods that occurred in the Czech Republic. The table indicates that mostly summer floods prevailed. Moreover, they usually consist of several waves. Flood intensity is defined as a return period of a watercourse's peak discharge rate in years (Q). The higher the value, the more serious and intense the flood. The table hence shows that in general, the Czech Republic experienced the most serious floods in years 1997 and 2002.

Weather extremes are not the sole cause of floods in the Czech Republic (and elsewhere). Other factors, such as urban development, river engineering, and agriculture amongst other human activities can influence river flood risk as well as climate variability. In flood-prone zones near riverbeds, people build housing, transport, industrial and other infrastructure. Although this topic has been reflected in the media due to a series of massive floods which occurred in Central Europe, affecting especially the eastern part of the Czech Republic (called Moravia) in 1997 and in the western Czech Republic (Bohemia) in 2002 and 2013, we still know little about the specific impacts of floods on vulnerable households and how they cope with them in this country.

The main aim of this work is to identify and analyze adaptation strategies of local households to impacts of extreme climatic phenomena, primarily flooding in the Czech Republic, and economical, demographical and other circumstances of adaptation. Our research focused on households living in 22 smaller municipalities, mainly in the Bečva River basin located in the northeastern part of the Czech Republic (see Fig. 5.1). The area was often affected by floods, which happened as a

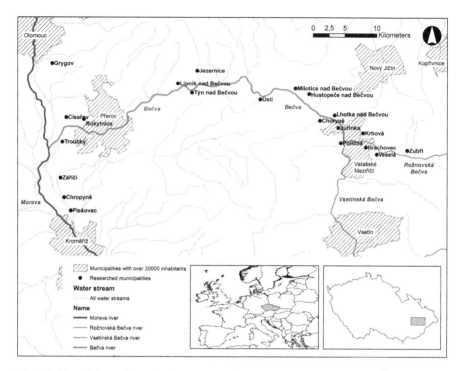

Fig. 5.1 Map of the study area in the north-east Czech Republic (Data source: ArcČR 500 digital database)

result of torrential or prolonged rainfall, with such climate extremes expected to worsen under climate change.

This chapter is structured into several parts: after this introductory section, section two is devoted to introducing theoretical framework of adaptation to climate extremes impacts, followed by the next section, which explains the data and used methods. The fourth section provides results of empirical research, and last section discusses these results and puts them into the wider perspective of adaptation strategies to climate extremes impacts, including key findings.

5.2 Climate Extremes Impacts and Adaptation

Changes within socio-economic systems incorporate changes in land use, municipal planning, regional development, increased risks as a consequence of building development in flood zones, and also a change in the perception of environmental risks, residing primarily in an underestimation of the potential flood risks. Technical measures have enabled people to move into originally inhospitable areas in river floodplains.

Increased fluctuations of extreme climatic and meteorological events have long been registered within the observed territory. For example, EM-DAT – International Disaster Database (2015) contains data about all types of disasters and their socio-economic impacts, gathered worldwide by the Centre for Research on the Epidemiology of Disasters (CRED). The database indicates floods as the most frequent disaster involving natural hazards in the Czech Republic, followed by extreme temperature events and storms (however, the critiques and limitations of the database ought to be considered as well).

As outlined in the introduction, in recent years the need to seek optimal methods of adaptation to ongoing or projected changes in climate has been emphasized at the international, as well as national and regional, level. Generally speaking, adaptation is seen as a common strategy of living organisms in adjusting themselves to changing environmental conditions including those related to climate. Climate change adaptation, as a theoretical construct with practical measures, is framed more specifically by IPCC authors as "an adjustment in natural or human systems in response to actual or expected climate stimuli to their effects, which moderates harm or exploits beneficial opportunities" (McCarthy et al. 2001; IPCC 2014).

In some natural systems, human intervention may facilitate adjustment to expected climate and its effects. According to Heffernan (2012), adaptation as a strategy to cope with climate change has not stayed, in contrast to mitigation strategies, at the centre of scientific focus so far. However, recently the reality of climate extremes in the form of extreme events like floods, droughts or heat waves (Lass et al. 2011) has forced researchers and policy-makers to explore ways to handle these extremes. Many regions face large natural and societal changes due to a combination of increased economic and residential assets in flood-prone areas, and increased societal vulnerability and lack of capability for handling climate extremes (Kreibich et al. 2005), climate change and its impacts.

If we are to focus on specific adaptation measures to a potential flood risk, Mechler and Kundzewicz (2010) differentiate between strategies of protection, adaptation and retreat.

A strategy of protection focuses primarily on ensuring a high degree of protection of the population and infrastructure against flood risks, residing in the implementation of "hard" structural measures (e.g. barriers, dams, relief channels or retention reservoirs). Despite the fact that these measures have contributed to reassuring the population and restoring their faith in protection against floods (e.g. Vaishar et al. 2002), the reality of flooding events has demonstrated that this strategy does not guarantee complete "protection" against the consequences of floods, and "protects" only to a certain extent (Etkin 1999; Fordham 1999).

A strategy of adaptation to floods or coexistence with floods represents the most sustainable strategy over the long-term, which reckons with a certain degree of flood risk, and seeks a combination of structural hard (technical) and soft measures, through a combination of preventive measures, including the rectification of damages and renewal.

The last and relatively radical strategy of retreat (sometimes termed "realignment") reckons with a withdrawal from flood-prone locations, thus a resettlement of the population or a relocation of economic activities. This strategy is problematic from a number of perspectives: People have developed floodplains to a large extent in the past, they remain attractive for further economic development or settlement, and people are relatively reluctant to abandon their place of residence. This topic is under-researched and within the central European context, low numbers of scholarly papers deal with the issue (for instance see Vikhrov et al. 2014; Stojanov et al. 2014b).

Those living in houses, in particular, are able to choose a construction solution that reduces or increases its potential resistance to the impact of climate extremes. The technical-architectural concept itself plays an important role. For example Botzen et al. (2013) conducted an economically focused study in the Netherlands, in which they determined respondents' willingness to pay for flood insurance in comparison with their willingness to pay for measures to reduce the flood risk in the form of constructing a raised ground floor of their house. The results demonstrated that approximately 52 % of respondents gave priority to a raised ground floor, thus they wished to resolve the problem rather than merely paying insurance. The authors believe that this approach is influenced by experiences hitherto in the Netherlands, where the number of flood events has increased markedly and people are aware, to a greater extent, of the risks. It is, however, necessary to state that only "willingness" was investigated and not actual measures.

Research into construction measures in order to reduce the flood risk to some extent and to improve the management of storm rainfall has been conducted for example by Kreibich et al. (2005). Kreibich et al. (2011), who use the term "precautionary measures," are amongst the authors who describe these measures in a relatively detailed manner, according to the economic costs in which it is necessary to invest. Amongst the low-cost measures are the gathering of information relating to precautionary measures, assistance of neighbours and relocation of at-risk possessions from the ground floor to safer locations. Medium investments include flood insurance, adaptation of the interior (e.g. floor replacement), securing of flood embankments and barriers. High-cost measures include the rebuilding the heating system in order to reduce the risk of it flooding; for example, relocating the boiler from the basement to a higher floor, by purchasing a mobile (demountable) boiler, or by removing under-floor heating on the ground floor. The highest investments are required by construction adjustments to the building using solid and water-resistant materials, raising the ground floor, sealing important parts of the house, fortifying the cellar and foundations of the building, or constructing small anti-flood walls on the surrounding lands.

5.3 Data and Methods

5.3.1 Methodology

We focused on a period when the studied households were affected by flooding from 1997 to 2012 in terms of what damage was caused, how they coped and how they are still adapting to floods. As part of analyzing the impacts of climate extremes, we examined development and changes in the population dynamics as a potential consequence of the changing flood regime. We asked whether the occupants perceive an increased incidence of floods and whether they are increasing in intensity.

The research questions have been set as follows: (i) *What types of adaptation measures are households taking with regard to the risk of occurrence of extreme weather?* and (ii) *Are characteristics such as out-migration and commuting for work different for individuals affected or not affected by floods?*

We focused on an analysis of the adaptive behavior of households to floods, which at the same time presents a conceptual framework for the research. It is based on a general differentiation of strategies for coping and adaptation, which are applied in the household. This framework thus incorporates wider ranges of potential behavior, beginning with the strategy of "resignation" or failure to take any measures. There follow simple, cheap, short-term measures (coping strategy), corresponding to minor adjustments within the house's interior. Measures which are more complex and more costly (adaptation strategies) are focused on the house's structure, primarily a raised ground floor or protection of the house against damp, along with the possible adjustment of its surrounding lands (IPCC 2012).

The last illustrated option is the strategy of migration, which does not mean that it is a measure on the highest level, but merely one of the possibilities of adaptive behavior. This concerns migration in the case of a significant change of the environment, a loss of dwelling because of an extreme hazard, or loss of livelihood (Stojanov et al. 2014a). In our case study, this would represent migration as a consequence of repeated flooding or other manifestations or impacts of extreme weather.

We applied descriptive statistics and statistical analysis through probit regression models dealing with relationships among factors influencing commuting. According to us, commuting may be viewed as the main migration factor available from survey responses. The probit model offers a consistent way regarding how to investigate the determinants influencing the decision to commute. We are interested in revealing the statistically significant variables that are able to discriminate between commuters and non-commuters. It is especially of great importance to find out the significant factors connected with adaptation strategies.

Empirical research was conducted by some authors of the chapter and interviewers who collected face-to-face respondents' answers. Collectors were trained by authors of the paper. We combined surveys among household residents located in the Bečva river basin and field research, based on the evaluation of the landscape, river dynamics and photo documentation of household adaptation measures.

5.3.2 Research Area and Sampling

The questionnaire investigation of the selected households took into account a number of fundamental conditions, one being that the respondents must dwell in a single family house, which means a detached house; i.e., a free-standing house, not any kind of apartment building. Nonetheless, several generations can live in the house together, such as a young family with elderly parents/grandparents. These residents must have had their permanent residence in this dwelling for at least 5 years. The basic circuits of the questionnaire were processed on the basis of a previous study and above all of the regional specifics of the region of the catchment basin of the Bečva, using a combination of closed and open questions. Households were selected as the fundamental research unit, which is a method frequently used by other authors from the perspective of New Economics of Labour Migration (see for instance Warner and Afifi 2014; Milan and Ruano 2014). Households tend to bear the consequences from the manifestations of climate change. Their potential to adapt is greatly overlooked in the context of flood impacts as well as in Czech research. In total, 605 household questionnaires (with 1739 residents) were obtained during the period from autumn 2012 to summer 2013 coming from 22 municipalities comprising 38,262 inhabitants.

The questionnaires were completed by going door-to-door, which means that the researcher asked a member of the household "at the door" of the house, where the respondent completed the questionnaire. This ensured the possibility of explaining those questions, which were not clear to respondents or of obtaining additional responses, personal opinions and other interesting information in connection with each specific response. The respondents were guaranteed anonymity and were not asked about the names of the members of the household. Households were selected geographically, so that we cover each risk zone proportionally (approximately one third of total household per each zone), and within each zone we strove to spread out houses equally, so they are not located too close to each other. Total number of households oscillated around 30 households/one municipality, depending on the distribution of the risk zones, the size of the municipality and the zone, and other factors.

Questionnaire interviewing of the members of the household was conducted in 22 small municipalities selected in advance. The municipalities were selected on the basis of a preliminary differentiation of the flood risk into zones of high, low and so-called no-risk, corresponding to the probability of flooding of the region in question within a certain time frame: Q 20 (high risk, corresponding to a probability of flooding once per 20 years), Q100 (low risk, corresponding to a probability of flooding once per 100 years) and so-called no risk (longer time dispersion, higher location above sea level, further from water sources). The flood territories for 20 and 100-year frequency of flooding are taken from the DIBAVOD database (DIBAVOD digital database 2015). The data on the water surfaces and water flows, similarly to the boundaries and positions of the individual municipalities, are taken from the database ArcČR (ArcČR 500 digital database 2015) and the Czech State Administration of Land Surveying and Cadastre.[1]

The division of each of the municipalities into three risk zones was applied in the research for which one third of households were addressed in each municipality in the high, low and so-called no-risk zones (see Fig. 5.2). The resulting sample oscil-

---

[1] For higher precision, all maps are shown in Křovák cartographic projection. The exception is the map of Europe, which is displayed in the standard World Geodetic System 1984.

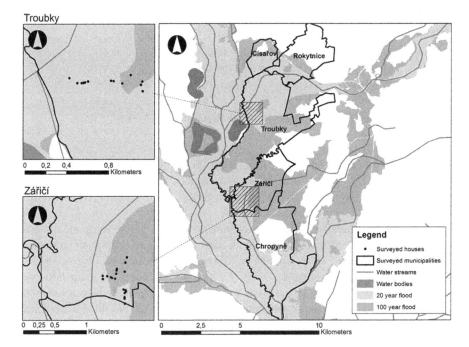

Fig. 5.2 Distribution of risk zones in individual municipalities in the example of the municipalities of Záříčí and Troubky (Source: Own data collected in the north-east Czech Republic, in 2012–2013)

lated around 30 questionnaires per municipality, in which this depended on the distribution of the risk zones, the size of the municipality, the willingness of the respondents to provide information and other factors. Because households were threatened not only by the River Bečva but also by its tributaries, for which maps of flood risk are not processed, during the course of the research, the risk zones on the tributaries of the River Bečva were determined according to the experiences of the households stated by the respondents in the questionnaire investigation.

5.4 Results

Our research amongst households living in the area of the Bečva River, Czech Republic shows both perception and evidence of an increasing intensity and frequency of the impacts of climate extremes in the form of floods over the last two decades. While from 1997 to 2004, the locals experienced two major floods, in the second half of the period under review, i.e. 2005–2012, there were four. While that indicates little from a statistical perspective, it has led to perceptions of increasing and worsening floods, often attributed (rightly or wrongly) to climate change. This perception is grounded by official data, as presented in the introduction.

Table 5.2 Distribution of houses in risk zones and their technical characteristics

Houses	Fired bricks	Non-fired bricks or combination	Flat land	Moderate slopes	Steep slopes	Total in risk zone
No risk	95 (58%)	55 (34%)	80 (49%)	67 (41%)	12 (7%)	163
Low risk	182 (68%)	87 (32%)	218 (81%)	46 (17%)	3 (1%)	269
High risk	118 (68%)	40 (23%)	154 (89%)	16 (9%)	2 (1%)	173

Source: Own data collected in the north-east Czech Republic, in 2012–2013

According to risk spatial preconditions (the official division into three risk zones) of the research, 73.2% (443) of households from our total sample experienced floods (corresponding with high- and low-risk zones), while 26.8% (162) of households had no experience with floods.

One of the key elements in the questionnaire was to detect how much and to what extent households experienced flood events, and what kind of adaptation strategies households undertook. The number of direct experiences with floods in the researched region depends on the zone. An absence of experience with flooding corresponds to a so-called no-risk area, one experience corresponds to a low risk area, and multiple experiences correspond to a high risk area. No flood experience was claimed by 162 households (26.8%), one flood by 268 households (44.3%), two floods by 141 households (23.3%) and three or more floods by 31 households (5.1%) over the period 1997–2012.

Regarding the technical characteristics and locations of houses, around 75% (452) were located on flat land, 21% (129) on moderate slopes and 3% (17) on steep slopes. Houses are made up of various building materials, 65% (395) of houses are built from fired brick, the most utilized building material in the Czech Republic, while 30% (181) were built of non-fired bricks or a combination of fired and non-fired bricks. The distribution of houses in risk zones depending on spatial dislocation and building material is characterized in Table 5.2. From the perspective, the houses from high risk zones are built on flat land (89%) and from fired bricks more frequently (68%) than houses constructed from non-fired bricks or a combination of both (23%). Fired bricks are more adaptable building material for houses in floods prone areas than non-fired bricks.

The presence of non-fired brick makes houses more vulnerable to flooding because water makes the bricks pulpy and destroys them. Non-fired bricks used to be employed traditionally so rather older houses contain non-fired bricks. They are not used contemporarily, due to their flood-prone characteristics and old-fashion style. A small proportion of houses, 4% (21), have a stone basement, which is also an old, traditional building material, but a highly effective adaptation measure against water flooding. Stone basements are usually elevated and are better protected, while they dry more effectively than other types. The rest of the houses were made from wood, ytong, breeze block and other materials.

A significant characteristic of adaptation measures is the height of the living level, the ground floor. Approximately half of the houses (46.9% or 284 houses) have their residential ground floor level approximately 1 m above the ground and

21.6% of the houses (128) have it even higher, which is the best variable for reducing flood risk. The average percentage of houses with a raised ground floor has fallen over the last 40–50 years from approximately 85% to almost 70%. With regard to the apparent higher occurrence of floods during recent years, this finding is surprising, despite the brief increase in the number of houses with raised ground floors following the floods of 1997. Moreover, we did not notice any significant differences among high-, low- and so-called no-risk zones; the proportion of houses with raised ground floors is similar. It indicates that owners of new houses, despite the flood risk, tend to follow "fashionable" or low-cost choices of houses without raised ground floors.

Czech legislation recommends, rather than demands, that building authorities elevate the ground floor for new houses in high flood risk zones. Current water legislation regulations included in the Czech Water Act No. 254/2001 request the incorporation of active floodplains into land use plans and forbid building new houses in active floodplain zones. In practice, and according to our personal experience from (not only) surveyed area, enforcement and monitoring of this issue are not strict, especially when political and development interests simply "delay" implementation of land use plans.

In terms of the extent of coping and adaptation measures in general and in response to floods, a certain percentage of households had already adopted some adaptation measures before the largest floods of 1997. In subsequent years, the number of implemented adaptation measures was only minimal, which can be seen from the fact that between 1997 and 2005, the inhabitants of the researched region were not confronted by any larger flooding, and the situation did not change markedly even after the 2010 floods. As regards to the number of exterior measures (reconstruction works provided on houses) per household, in 58% of households, no adaptation measures were implemented, 24.3% of households implemented only one type of measure, 12.2% of households implemented two types only and 5.5% of households adopted three or more types of adaptation measures.

We recognized three most frequently used adaptation measures:

- Hydro-isolation of the house walls. This is not generally considered to be a special measure against river floods. Rather, it is a basic way to avoid the house getting damp from wet ground. In total 25% of households living in high risk zones used this adaptation measure (for details about other zones see Table 5.3).
- Hydro-isolation combined with drainage. This is a higher level of house protection against soaking and small-scale flooding from rainfall. The drainage system around the house is more sophisticated, consisting of retention pipes and gravel around the house to prevent water and moisture from infiltrating. Hydro-isolation could partially help to protect a house from rainfall flooding and enables the house to dry faster. Combination of both techniques was used by 40% households living in high risk zones.
- Water management of the plot and terrain and vegetation adjustments. These are two options that are difficult to distinguish, but they include some adjustments of plot and terrain around the house such as self-made ditches, walls, and drainage

Table 5.3 Distribution of most frequent adaptation measures according to zones

Adaptation measures	No risk	Low risk	High risk	Total
Hydro-isolation of the house walls	9 (6%)	58 (22%)	43 (25%)	110
Hydro-isolation combined with drainage	30 (18%)	73 (27%)	68 (40%)	171
Water management of the plot and terrain and vegetation adjustments	14 (9%)	40 (15%)	42 (24%)	96

Source: Own data collected in the north-east Czech Republic, in 2012–2013
Note: Percentages are computed as a ratio to all households in corresponding zones

channels. Even though useful, an expensive approach is a professional system of underground local water drainage designed to increase the soil retention capacity of water near the house. This adaptation measure was used by 24% households living in high risk zones.

Table 5.3 summarizes the distribution of adaptation measures according to risk zones. Hydro-isolation combined with drainage is one of the most used measures regardless of the particular risk zones. More than half of the households living in the high risk zones did not apply any of the adaptation measures discussed above.

Households tend to prefer simple and cheap solutions such as moving possessions upstairs or getting some mobile barriers, compared with a lower number of households that changed the floor material for a more water-resistant one. The most common application of hydro-isolation can be explained in that it is a standard and basic way of avoiding dampness in the house from wet ground, instead of being a special flood risk reduction measure. Hydro-isolation combined with drainage provides a higher level of protection from damp and flooding.

Some respondents mentioned that they applied other measures; for example, build a small lifting bridge, applying special plaster on walls or leaving walls without plaster at all (some people found it useless due to frequent floods). One respondent even built a wall against flooding. Besides the mentioned measures, one respondent mentioned he bought a pump in case of flooding and a mobile boiler, which could be protected against flooding.

When we compare adaptation measures taken by risk zone, households in high and low risk areas account for approximately 80% of all measures taken by all households. Those households in so-called no risk zones take few measures. There is some visible pattern of action based on experience and flood risk awareness. The higher the probability of flood where a house is located, the more adaptation measures the household tends to apply. These results differ from Siegrist and Gutscher (2006) who found that, for Switzerland, flood risk perception or flood experience did not significantly support application of flood damage prevention measures. Conversely, Miceli et al. (2008) conducted research in Italy, supporting the hypothesis concerning the significant role of perception in supporting adaptation measures. They found that the higher the flood risk perception, the higher the number of household that applied some measures to reduce flood risk.

Only two respondents confirmed moving away from the flood risk site to a safer place thanks to municipal support. Our data showed that insurance companies

Table 5.4 Probit model – all individuals (employees)

	Coefficient estimate	Standard error	p-value
Intercept	−0.9487	0.6216	0.1269
nexp_2	−0.3976	0.2483	0.1093
adap_sum_0610	0.3000	0.1817	0.0987
adap_sum_97	−0.1355	0.0680	0.0464
adap_sum_9706	−0.1771	0.0543	0.0011
risk_2	0.2029	0.1410	0.1501
risk_3	0.7384	0.2643	0.0052
Data and model characteristics			
No. of employees	686	Non-commuting	276
Region fixed effect	included	Commuting	410
Association of predicted probabilities			
Percent concordant	72.4	Percent discordant	27.5

Source: Own data collected in the north-east Czech Republic, in 2012–2013
Definition of variables:
nexp_2 – Dummy variable, experience with floods in 2 years at least = 1, otherwise 0
adap_sum_97 – Sum of adaptation measures (internal and external) before 1997 (0–7; 4 external, 3 internal)
adap_sum_9706 – Sum of adaptation measures (internal and external) in the period 1997–2006 (0–7; 4 external, 3 internal)
adap_sum_0610 – Sum of adaptation measures (internal and external) in the period 2006–2010 (0–7; 4 external, 3 internal)
risk_2 – Dummy variable; household living in low risk area = 1, otherwise 0
risk_3 – Dummy variable; household living in high risk area = 1, otherwise 0

tended not to be willing to reimburse more than 50–60 % of losses and some houses were not eligible for insurance compensation. That meant many affected people had few opportunities for resettlement, due to lack of funds, even if they wished to relocate.

Another population dynamics adaptation strategy is commuting for work. Factors influencing decisions of individuals about commuting to work were investigated using the probit regression model. Probit model was used for analyzing commuting patterns only, i.e. not for other types of adaptation. Table 5.4 presents statistically significant factors, which could be linked to the experiences with floods and realization of adaptation measures. As for the other significant factors (not presented here), individual characteristics (age, education, earnings) have been controlled for as well. From the results of the parameter estimates of the probit model (see Table 5.4), there is an evident relationship between experiences with floods, realization of adaptation measures and commuting. Our sample consists of all employed individuals and we have excluded students, children and retirees. This sample comprises 686 employed individuals of which 410 are commuters and 276 are non-commuters.

Respondents who are subject to a second flood (nexp_2) are less likely to commute in comparison to those unaffected by floods or affected by only one flood. Respondents implementing more adaptation measures in the period 2006–2009 (adap_sum_0610) are more likely to commute. Conversely, respondents who implemented more adaptation measures prior to the year 1997 (adap_sum_97) and in the period of 1997–2005 (adap_sum_9706) are less likely to commute. From this point of view, commuting may be viewed as an adaptation strategy substituting local or physical measures.

Moreover, respondents living in the low risk zones (risk_2) are more likely commute than the respondents from so-called no-risk zones and respondents living in the high risk zones (risk_3) are more likely commute than the respondents from so-called no-risk zones. This effect is more statistically significant than the one of low risk zones.

From other probit models based on comparison of individuals (employees) affected by one flood, at least two floods or unaffected by any floods, other interesting trends emerge. For instance, respondents living in the high risk zones are more likely commute than the respondents from so-called no-risk or low risk zones. Respondents affected by at least two floods or unaffected by any floods for which the losses incurred after the first flood exceed CZK 50,000 tend to have lower probability for commuting. Finally, respondents affected by one flood or unaffected by any floods perceiving the flood preparedness of their households as very good tend to be less likely to commute than those appraising it as very bad.

Further, we found that commuters on average have higher earnings than non-commuters. The relationship between commuting and exposure to floods is non-linear because individuals commute more after the first flood and less after the second flood. Our explanation for the difference is in unfavorable economic conditions of the target region. In particular, men aged between 30 and 50 years old who had finished secondary school tend to move away from risky regions after the first flood. That means that groups of people who moved away consisted of more active people who originally commuted more often to work than those who decided to stay in the given region. They also left behind their houses, which has been detrimental to community development.

5.5 Concluding Remarks

Our research among households living in the area of the Bečva River basin in the Czech Republic shows a perceived increasing intensity and frequency of climate extremes impacts from floods within last two decades. Many residents felt that they were experiencing a trend of more frequently alternating extreme rainfall and drought. Our research also indicated, in addition to the risk of flooding from the Bečva River, a large number of households in the foothills of the Beskydy Mountains have experienced problems with flash floods from its tributaries.

We recognized a set of household-level adaptation strategies outside houses (such as terraces, elevated ground floor constructions, and hydro-isolation from external water) and highlighted economic costs connected with floods, insurance strategies, and population dynamics. The strategy of large-scale migration from the floodplains of rivers has not been applied, even in high-risk zones. Our research did not show an extensive increase in adaptation measures adopted at the household level; households have a tendency to repair damage rather than to implement costly adaptation measures.

So far, there is no support (e.g. subsidies or tax breaks) in the Czech Republic for policies that would support households affected by flooding and trying to implement adaptation measures including migration. In the future, increasing emphasis should be placed on more comprehensive and integrated adaptation solutions along with communication and consultation with those affected. Solutions should not only be based on flood risk reduction, but also on connections to land use planning, water resource management, urban and rural development, and other relevant and rather complex measures reflecting projected future climate trends.

The level of transferability or generalizability from the Czech case study is difficult to determine, but we have indicated above where our results do and do not match literature from other locations. Certainly, the lessons from our work indicate that migration as an adaptation measure ought to be considered in any similar circumstances, although interest and ability to implement migration by either people themselves or the authorities will vary with context. It is clear that possibilities emerge to support people who would wish to migrate to reduce their risk, but cannot for financial reasons.

However, our research did not show extensive out-migration from floods prone areas, we recognized population dynamics in the form of commuting for a job as one of risk management strategy of locals. The commuting for work needs to be explored further in the Czech Republic and elsewhere to determine whether or not it is a viable adaptation strategy and the level of affinity which people have for it; that is, they might be adopting migration as an adaptation strategy because they feel that they have to, rather than because they feel that they want to. Consequently, population dynamics (and desires for moving or not) of different forms can play important roles in adaptation to climate extremes impacts, but care needs to be taken to ensure that decisions are not forced on unwilling people. Instead, people deserve the dignity of choice for population dynamics.

Acknowledgment We would like to express our gratitude to the respondents who shared their information and the interviewers who patiently collected their answers.

References

ArcČR 500 Digital Database. (2015). ArcData Praha. http://www.arcdata.cz/produkty-a-sluzby/geograficka-data/arccr-500/. Accessed 16 Apr 2015.

Bardsley, D. K., & Hugo, G. J. (2010). Migration and climate change: Examining thresholds of change to guide effective adaptation decision-making. *Population and Environment, 32*(2–3), 238–262.

Black, R., Bennett, S. R. G., Thomas, S. M., & Beddington, J. R. (2011). Migration as adaptation. *Nature, 478*, 447–479.

Botzen, W. J. W., Aerts, J. C. J. H., & van den Bergh, J. C. J. (2013). Individual preferences for reducing flood risk to near zero through elevation. *Mitigation and Adaptation Strategies for Global Change, 18*(2), 229–244.

CHMI. (2013). *Evaluation of the most serious floods in the Czech Republic:* 1997, 2002, 2006, 2009, 2010, 2013. http://voda.chmi.cz/pov/index.html. Accessed 16 Apr 2015.

DIBAVOD Digital Database. (2015). T. G. Masaryk Water Research Institute. http://www.dibavod.cz/. Accessed 16 Apr 2015.

Duží, B., Vikhrov, D., Kelman, I., Stojanov, R. & Jakubínský, J. (2015a). Household flood risk reduction in the Czech Republic. *Mitigation and Adaptation Strategies for Global Change, 20*(4), 499–504.

Duží, B., Vikhrov, D., Kelman, I., Stojanov, R. & Juřička, D. (2015b). Household measures for river flood risk reduction in the Czech Republic. *Journal of Flood Risk Management.* doi:10.1111/jfr3.12132. online.

EEA. (2012). *Climate change. impacts and vulnerability in Europe 2012.* EEA report No. 12. Copenhagen. 2012.

EM-DAT – International Disaster Database. (2015). *Centre for research on the epidemiology of disasters.* http://www.emdat.be/database/. Accessed 16 Apr 2015

Etkin, D. (1999). Risk transference and related trends: Driving forces towards more mega-disasters. *Environmental Hazards, 1*, 69–75.

Fordham, M. (1999). Participatory planning for flood mitigation: Models and approaches. *Australian Journal of Emergency Management, 13*(4), 27–34.

Heffernan, O. (2012). No going back. *Nature, 491*, 659–661.

Intergovernmental Panel on Climate Change (IPCC). (2012). *Managing the risks of extreme events and disasters to advance climate change adaptation* (A special report of Working Groups I and II of the Intergovernmental Panel on Climate Change). Geneva: Intergovernmental Panel on Climate Change.

Intergovernmental Panel on Climate Change (IPCC). (2014). Summary for policymakers. In *Climate change 2014: Impacts, adaptation, and vulnerability.* Part A: Global and Sectoral Aspects. Contribution of Working Group II to the Fifth Assessment Report of the Intergovernmental Panel on Climate Change. Cambridge: Cambridge University Press.

Jongman, B., et al. (2014). Increasing stress on disaster-risk finance due to large floods. *Nature Climate Change, 4*, 264–268.

Kelman, I., Stojanov, R., Khan, S., Alvarez Gila, O., Duží, B., & Vikhrov, D. (2015). Islander mobilities: Any change from climate change? *International Journal of Global Warming, 8*(4), 584–602.

Kreibich, H., Thieken, A. H., Petrow, T., Müller, M. & Merz, B. (2005). Flood loss reduction of private households due to building precautionary measures – lessons learned from the Elbe flood in August 2002. *Natural Hazard and Earth System Sciences, 5*(1), 117–126.

Kreibich, H., Seifert, I., Thieken, A. H., Lindquist, E., Wagner, W. & Merz, B. (2011). Recent changes in flood preparedness of private households and business in Germany. *Regional Environmental Change, 11*(1), 59–71.

Lass, W., Haas, A., Hinkel, J., & Jaeger, C. (2011). Avoiding the avoidable: Towards a european heat waves risk governance. *International Journal of Disaster Risk Science, 1*, 1–14.

McCarthy, J. J., et al. (2001). *Climate change 2001. Impacts, adaptation, and vulnerability* (Contribution of Working Group II to the third assessment report of the Intergovernmental Panel on Climate Change). Cambridge: Cambridge University Press.

McLeman, R., & Hunter, L. M. (2010). Migration in the context of vulnerability and adaptation to climate change: Insights from analogues. *WIREs Climate Change, 1*, 450–461.

Mechler, R., & Kundzewicz, Z. W. (2010). Assessing adaptation to extreme weather events in Europe – Editorial. *Mitigation and Adaptation Strategies to Global Change, 15*, 611–620.

Miceli, R., Sotgiu, I., & Settanni, M. (2008). Disaster preparedness and perception of flood risk: A study in an alpine valley in Italy. *Journal of Environmental Psychology, 28*(2), 164–173.

Milan, A., & Ruano, S. (2014). Rainfall variability, food insecurity and migration in Cabricán, Guatemala. *Climate and Development, 6*(1), 61–68.
Neumann, K., & Hilderink, H. (2015). Opportunities and challenges for investigating the environment-migration nexus. *Human Ecology, 43*, 309–322.
Oliver-Smith, A. (2012). Debating environmental migration: Society, nature and population displacement in climate change. *Journal of International Development, 24*(8), 1058–1070.
Plummer, R., & Baird, J. (2013). Adaptive co- management for climate change adaptation: Consideration for the Barents Region. *Sustainability, 5*, 629–642. doi:10.3390/su5020629.
Reid, H., Alam, M., Berger, R., Cannon, T., Huq, S. & Milligan, A. (2009). *Participatory learning and action: Community-based adaptation to climate change*. London: International Institute for Environment and Development (IIED).
Siegrist, M., & Gutscher, H. (2006). Flooding risks: A comparison of lay people's perceptions and expert's assessments in Switzerland. *Risk Analysis, 26*(4), 971–979.
Stojanov, R., Kelman, I., Shen, S., Duží, B., Upadhyay, H., Vikhrov, D., Lingraj, G. J., & Mishra, A. (2014a). Contextualising typologies of environmentally induced population movement. *Disaster Prevention and Management, 23*(5), 508–523.
Stojanov, R., Kelman, I., Martin, M., Vikhrov, D., Kniveton, D., & Duží, B. (2014b). *Migration as adaptation? Population dynamics in the age of climate variability*. Brno: Global Change Research Centre, Academy of the Sciences of the Czech Republic.
Tacoli, C. (2009). Crisis or adaptation? Migration and climate change in a context of high mobility. *Environment and Urbanization, 21*(2), 513–525.
Vaishar, A., Hrádek, M., Kallabová, E., Kirchner, K., Klímová, A., Klusáček, P., Kolibová, B., Lacina, J., Munzar, J., Střítežská, Š. & Ondráček, S. (2002). *Krajina. lidé a povodně v povodí řeky Moravy: regionálně geografická studie*. Brno: Regiograph. 131 p.
Vikhrov, D., Stojanov, R., Duží, B. & Juřička, D. (2014). Commuting patterns of Czech households exposed to flood risk from the Becva river. *Environmental Hazards, 13*(1), 58–72.
Walker, W. E., Haasnoot, M., & Kwakkel, J. H. (2013). Adapt or perish: A review for planning approaches for adaptation under deep uncertainty. *Sustainability, 5*, 955–979. doi:10.3390/su5030955.
Wamsler, C., & Brink, E. (2014). Planning for climatic extremes and variability: A review of Swedish municipalities' adaptation responses. *Sustainability, 6*, 1359–1385. doi:10.3390/su6031359.
Warner, K., & Afifi, T. (2014). Where the rain falls: Evidence from 8 countries on how vulnerable households use migration to manage the risk of rainfall variability and food insecurity. *Climate and Development, 6*(1), 1–17.

Chapter 6
Moving Beyond the Focus on Environmental Migration Towards Recognizing the Normality of Translocal Lives: Insights from Bangladesh

Benjamin Etzold and Bishawjit Mallick

6.1 Introduction

Bangladesh is a country of 160 million people that is situated right at the heart of the Ganges-Brahmaputra-Delta in South Asia. It is a densely populated (1019 people/km^2) low-income (GDP of US$ 150 Billion) country. One third of the population lives below the poverty line (UNDP 2014). Many of the poor are highly mobile as they regularly (have to) search for livelihood opportunities in other places. There is mounting evidence that natural hazards such as floods, cyclones, and droughts increase in frequency in Bangladesh and that creeping processes such as river erosion, sea-level rise, and salinity ingress are going to continue unabated. In the future, the already existing high variability of rainfall is likely to be further accentuated. Together, these changes add to persisting patterns of stress on ecosystems and to land degradation and thereby impact rural livelihoods, agricultural productivity, and people's food security (IPCC 2007, 2014a). Both extreme events and subtle climatic changes challenge people's capacities to cope with and recover from incurred damages and to adapt to a changing environment in the long run.

Employing a social vulnerability perspective, many scholars investigate how people live in contexts of risks and uncertainties. They investigate people's exposure and sensitivity to natural hazards and environmental crises, highlight structural causes of people's vulnerability, and pay particular attention to their adaptive

---

B. Etzold (✉)
Department of Geography, University of Bonn,
Meckenheimer Allee 166, 53115 Bonn, Germany
e-mail: etzold@giub.uni-bonn.de

B. Mallick
Institute of Regional Science (IfR) at the Karlsruhe of Technology (KIT),
Reinhard-Baumeister-Platz 1, 76131 Karlsruhe, Germany
e-mail: bishawjit.mallick@kit.edu

© Springer International Publishing Switzerland 2016
A. Milan et al. (eds.), *Migration, Risk Management and Climate Change:
Evidence and Policy Responses*, Global Migration Issues 6,
DOI 10.1007/978-3-319-42922-9_6

capacities (Bohle 2007; Wisner et al. 2004). The absolute extent of a flood, the destructive power of a cyclone, or the length of a drought might then be less important than 'root causes' of poverty and the actions and interventions that shape people's vulnerability. Seen in this context, migration can serve as a 'risk management strategy' for vulnerable people, who seek to secure and diversify their livelihoods in the longer run, and who act in response to everyday insecurities and weather-related risks.

In the popular discourse about climate change and adaptation in Bangladesh, migration is most often framed in a negative light. People have to 'flee' from natural hazards or they are 'displaced' in the wake of disasters. Thereafter, they have to cope with the effects of both the disaster and their displacement. Migration is then largely seen as a reaction and migrants as more or less passive 'victims' of environmental change. While numerous studies seem to prove this observation, others increasingly recognize migrants as people with agency and as drivers of change. Mobility opens up livelihood opportunities in multiple places. By living 'translocal lives', households with migrant family members might be even more resilient to natural hazards, subtle environmental changes and economic turbulences. Seen in this light, migration is more than a strategy of 'risk management' or 'reactive climate change adaptation'. It is a question of livelihood choices, human rights and freedoms of movement.

It is argued in this chapter that a social vulnerability perspective helps us to understand 'environmentally-induced migration' and people's translocal lives in Bangladesh in a more differentiated way. After a conceptual section with definitions of key terms and a brief introduction into migration patterns inside and beyond Bangladesh, it presents an overview of the contemporary debate about weather-related risks and migration in Bangladesh. It also includes a brief discussion of popular discourses on climate change and migration and political responses to the portrayed crises. Most importantly, it contextualizes migration and translocality as a rule rather than as a 'state of exception' in the academic discourse.

6.2 Environmental Migration, Vulnerability and Translocality

In the academic debate about climate change, migration is often discussed as a coping strategy against rapid-onset natural hazards and as an adaption to slow-onset processes. If people leave a place because their livelihoods have been negatively affected by natural hazards or environmental changes, one might speak of "environmentally-induced migration" (see McLeman and Smit 2006; Piguet et al. 2011b; Warner et al. 2010 for an introduction to the debate and its contested terminology). The International Organization for Migration (IOM) defined environmental migrants as "persons or groups of persons, who for reasons of sudden or progressive changes in the environment that adversely affect their living or living conditions, are obliged to leave their habitual homes, or choose to do so, either temporarily or permanently, and who move either within their country or abroad" (IOM 2010: 4). These migrants are thus portrayed as vulnerable people who react to environmental transformations.

Social scientists and climate scientists often mean different things when they use the term vulnerability. Most climate scientists view vulnerability in terms of the likelihood of occurrence and impacts of weather and climate related events. The IPCC, for instance, defines vulnerability as the "propensity or predisposition to be adversely affected" (IPCC 2014b: 6). According to Adger (2006: 268), vulnerability is "the state of susceptibility to harm from exposure to stresses associated with environmental and social change and from the absence of capacity to adapt". This definition contains some key components: *shocks* or *stresses* that pose a challenge to people; *exposure* in terms of the common risk to encounter the aforementioned stresses; *sensitivity* in terms of the individual risk of being susceptible to that stress; and *adaptive capacity* in terms of the individual and collective ability to anticipate stress, cope with its short-term effects and adapt to its longer term impacts (Adger 2006; Füssel 2007; Smit and Wandel 2006). A social vulnerability perspective pays particular attention to societal rules, power relations, social practices, and human agency in order to understand how people's exposures, sensitivities and adaptive capacities are shaped (Bohle 2007; Wisner et al. 2004). It conceives of vulnerability as being "embedded in social and environmental arenas, where human security, freedoms, and human rights are struggled for, negotiated, lost and won" (Bohle 2007: 9).

From a social vulnerability perspective, it might thus be misleading to single out 'environmental factors' or even just 'weather-related risks' as the key drivers of people's vulnerability, and hence consider migration as a reactive strategy of people who are facing these risks. Instead, societal relations, political economies and other structural transformations are of equal importance as drivers of both people's vulnerability and their mobility (Black et al. 2011; Foresight 2011). As follows, migration patterns must be considered as multi-causal social processes that cannot be fully understood by solely looking at the 'push factors' that force people to leave their home region in reaction to structural conditions or an omnipresent threat, and the 'pull factors' that entice some to go to another place to optimize their life chances (even though listing this factors might be helpful; see Mallick and Vogt 2012; Arsenault et al. 2015). Structural interdependencies and migration systems that emerged over time; personal networks that link different places; people who provide migration services to others; state actions as well as legal, social and economic barriers that hinder mobility; perceptions of the living conditions elsewhere; and cultural meanings of being mobile or staying put; all these aspects influence people's migrations as well as societal discourses about mobility and mobile people (Samers 2010). Moreover, recent research on translocality indicates that migrants do not necessarily depart from a place of origin and permanently settle and integrate at a place of destination, which indicates that they 'leave their home behind'. Most mobile people rather remain situated in one socio-spatial unit – a "translocal social field" –

which includes quite different social settings and stretches over multiple places, and which is (re)produced by personal relations and social networks. Migrants' everyday life and their families' translocal livelihoods are characterised by their experience of mobility, their "simultaneous embeddedness" in, and their networks across specific local places (Brickel and Datta 2011; Greiner and Sakdapolrak 2013; Etzold 2016; Levitt and Glick Schiller 2004; Thieme 2007). Likewise, social vulnerability is not only experienced in one place, but in multiple places and social contexts. In the following, we sketch out some basic structural trends that should not be forgotten in the contemporary debate about 'environmentally-induced migration' in Bangladesh. We argue that migration is more than a reactive coping strategy by which people are trying to reduce their vulnerability and more than a proactive adaptation strategy by which actors seek to open up and make use of livelihood opportunities at multiple places (Mallick and Siddiqui 2015). In Bangladesh, migration has simply become a part of people's everyday life. Translocal lives have become the reality, and cannot be considered as an exception.

6.3 Contextualizing Current Migration Patterns in Bangladesh

Current migration patterns in the context of climate change or globalization cannot be understood without bearing in mind broader structural changes over time. People have been moving throughout the Bengal delta for centuries, but the majority of the population has been sedentary and live in rural villages. Independence from colonial rule and the partition of British India in 1947 triggered the displacement of hundreds of thousands of Hindus and Muslims between India and Pakistan. In the 1960s, the population of East Pakistan increased quickly. With industrialization and economic growth, villagers began to migrate to cities for alternative livelihoods. After a 9 month long civil war, during which up to three million people lost their lives and ten million fled to India, Bangladesh gained its independence in 1971. In 1974, severe floods destroyed harvests and grain stocks, whilst market speculation led to a rapid increase of the price of rice. As a result, tens of thousands of people died of hunger and even more people were driven to urban areas in search of food and income. After the famine, people continued to move from rural to urban areas, spurring urbanization and in particular the growth of the national capital Dhaka (Afsar 2005; van Schendel 2009).

Since the 1980s, migration patterns in and from Bangladesh were affected by two major trends. On the one hand, rising oil prices and thus an increasing need for cheap labor power in the Middle East and in Southeast Asia led to the growth of international labor migration from Bangladesh. The number of laborers who left for work in the Gulf states increased tenfold from 25,000 in 1980 to more than 250,000 in 2010 (Rahman 2012; Siddiqui 2005). Overseas workers from Bangladesh were integrated in existing migration systems and played their role in the global division of labor. The outflow of migrants from Bangladesh has contributed significantly to social, economic, and political change in the nation and led to formation of a transnational social field that connects the Bangladeshi diaspora, temporary international labor migrants, and the society back home (Dannecker 2005; Gardner 1995). On the other hand, Bangladesh started to produce and export textiles in the early 1980s, which gradually changed its role in the global economy. In 1985, roughly 120,000 people worked in 380 garment factories, while it was around 1.6 million workers in 3200 factories in 2000, and even four million workers in 4200 factories in 2014 (BGMEA n.d.). The garment factories are predominantly situated in and around Dhaka, which fueled its economic growth and further enhanced internal migration flows. The megacity's industrial boom also led to social transformations as young rural women, who did not migrate in large numbers before, gained access to livelihood opportunities in urban factories (Afsar 2005; Dannecker 2002; Haan et al. 2000; Salway et al. 2003). In between 1980 and 2015, Dhaka's population increased from 3.3 to 17.6 million people, largely due to in-migration (UN/DESA 2014). In the nation's young history, Dhaka has always been the most important destination for labor migrants. These migrants are functionally integrated in the global economy.

In 2015, Bangladesh had an estimated population of 160 million people. Two Third of the population still lives in rural areas; one Third resides in urban areas (UN/DESA 2014). However, due to prevailing poverty and food insecurity in many parts of the country, regular disruptions of rural livelihoods by natural hazards, more diverse economies in cities, centralistic educational structures, and improved transportation networks, the life of many families has become increasingly mobile, and even translocal. The 2011 population census revealed that 12 % of all Bangladeshi households have migrants in their family. Roughly 13.5 Million people have left the district in which they were born during their lifetime, which is 10 % of the Bangladeshi population. Of these internal migrants, 44 % have moved from rural to urban areas, another 43 % from rural to rural areas, 9 % from one city to another, and only 4 % from urban to rural areas. Although migration is an everyday practice in Bangladesh, nowadays, with 90 % the vast majority of the population has not (yet) been mobile themselves (BBS 2012).

6.4 Environmentally-Induced Migration in Bangladesh

In the last decade, besides international labor migration to the Gulf states and rural-urban migration, a third trend has shaped the debate about migration in and from Bangladesh: extreme weather events and climate change. In general, there is a rapidly growing body of literature on the relationships between natural hazards and weather-related risks on the one hand, and migration and displacement on the other hand (Afifi and Jäger 2010; Black et al. 2011; Foresight 2011; Piguet et al. 2011a; Warner and Afifi 2014; Warner et al. 2010). Bangladesh is one of the most frequently studied countries. To sort this field, one might best distinguish between different environmental drivers of livelihood changes and human mobility.

The report "Assessing the evidence: Environment, climate change and migration in Bangladesh" by the International Organization for Migration (IOM 2010) used three categories for its assessment: mobility in the context of sudden-onset events, migration in relation to slow-onset processes, and the 'cascade' effects between environmental degradation, urbanization, human security and migration. In the following we will look at the literature and provide some examples in the first two fields.

6.4.1 Displacement, Migration, Immobility and Return in the Context of Natural Hazards

For Bangladesh, ample empirical evidence exists on the effects of climate-related *sudden-onset events* or *natural hazards*, such as floods or tropical cyclones, on people's mobility (Black et al. 2013; IOM 2010; Mallick and Etzold 2015; Mallick and Vogt 2012; Paul 2005; Penning-Rowsell et al. 2013; Poncolet et al. 2010; Warner et al. 2009; World Bank 2010).

Floods are a fact of everyday life for many Bangladeshis. Bangladeshis are used to 'living with floods'. They have adapted to floods by raising their houses on plinths and adjusting their farming systems – in rural areas the seasonal rhythm of labor is adjusted to the agricultural seasons and the periods when floods are likely to occur. Nonetheless, excessive floods can become a natural disaster. In the last 25 years, Bangladesh has experienced 6 severe floods causing 9600 deaths. The floods of 1987, for instance, led not only to the loss of 2000–6000 lives, and the destruction of public infrastructure, private houses, and agricultural land, but also to temporary displacement of 45 million people within Bangladesh (IOM 2010). The challenges of displacement during a flood and the benefits of migration patterns that emerge before and after floods need to be seen in the context of the coping strategies that are locally available for flood-affected people. Microcredits and micro-insurance might help people to overcome negative long-term effects of both flooding and displacement. But they also might create new dependencies (Akter and Fatema 2015; Gehlich-Shillabeer 2008).

Tropical storms can hit local communities hard and have the power to destroy properties, livestock, and people's lives. Yet, the mid- and long-term effects on food production, rural livelihoods, and mobility patterns can be even more severe (Kartiki and Kartiki 2011; Mallick and Vogt 2014). Paul (2005) provides evidence that micro-scale disasters such as a *tornado* that occurred in north-central Bangladesh led to migration. *Tropical cyclones* that are usually accompanied by high winds and storm-surges hit Bangladesh every 3 years on average. Since 1877, a total of 160 cyclones have formed in the Bay of Bengal. The Bangladesh Meteorological Department reported that more than one million people were killed in those storms (Mallick and Vogt 2012). The more recent cyclones Sidr (in 2007), Alia (in 2009) and Mohasen (in 2011) affected millions of people in Bangladesh, but due to improvements in early warning systems and disaster emergency help, the death toll was far smaller than in the great storms of 1970 and 1991. Cyclones can affect migration patterns through different pathways. On the one hand, many survivors have been only temporarily and locally displaced from their homes and quickly returned. On the other hand, thousands of people lost their homes and livelihoods and migrated to cities permanently in search of shelter, employment, and secure lives (Findlay and Geddes 2011; IOM 2010; Poncolet et al. 2010).

Textbox 6.1: Cyclone Induced Population Displacement in Southwest Coastal Bangladesh

In a detailed analysis of rural livelihoods and people's post-disaster coping strategies carried out in Bangladesh's coastal zone, Mallick and Vogt (2012) found that one third of almost 300 households had 1 or more members – mostly men – who migrated temporarily to nearby cities. Because cyclone Alia also destroyed the employment opportunities in their villages, these individuals left home right after the cyclone to search for alternative sources of income. Seventeen percent of the respondents sold some assets in order to cope with the adverse situation after the cyclone. Two Third of them sold their cattle and other livestock, not only because the needed cash, but also because of a lack of fodder and adequate animal shelters. Sixty percent sold timber or wood plants. Six percent sold ornaments or other household assets like TV, mobile phones, radios, etc. None of them sold their land. Further adaptive strategies were also investigated. It showed that 29 % of those households that sold some assets after the cyclone had also changed their major sources of income. Only 3 % reported that changes in income sources had improved their economic solvency. Among those who sold out their assets, 82 % reported that they had to take credit (ranging from 2000 to 25,000 BDT) from different sources to cope with the effects of the cyclone; 69 % received loans from NGOs, 13 % from local money lenders, 12 % from relatives and only 6 % from Banks.

The survey showed that 29 % of the respondents needed to migrate or commute to nearer cities to overcome the adverse situation, even after selling out their resources. They left their homes immediately after the emergency relief works had stopped; i.e. 4 weeks after the cyclone. The population displacement that occurred immediately after the cyclone event does not show any causal relationship with the households' socio-economic conditions and the state's and NGO's intervention programs. The short-term recovery support has not been relevant for people's decision to migrate (or not), but rather for immediate recovery and reconstruction activities. The assistance of local politicians or other influential people did, however, play a decisive role in the migration-decisions of the cyclone-affected people. Nineteen percent of the respondents argued that they had to migrate, because they were not supported

(continued)

Textbox 6.1 (continued)
by their local politicians. When the post-cyclone reconstruction activities under a state-run Food for Work (FFW) program had started, none of them were included in the lists of people who are eligible for such a program. As follows, they then left their home villages to seek for alternative sources of income. Inclusion in the Food for Work activities has been a key indicator of connectedness to local politicians. Access to such networks of support was a decisive factor for the survivors' access to the local post-cyclone labor market. It also showed that members of those households, which had prior knowledge about possible destinations and which had translocal networks that could facilitate access to work at the respective destination, were more likely inclined to migrate. Most of the post-disaster migrants (78 %) moved to bigger cities in Bangladesh's coastal zone, namely Satkhira, Khulna and Bagerhat, where they saw possibilities to get at least a job as rickshaw driver or day laborer (Mallick 2015; Mallick and Vogt 2012, 2014).

This case study shows that landless, resource-constrained and politically marginalized people in the exposed areas are often not only most severely affected by natural disasters, they also face more difficulties in their coping actions (for other examples see Islam et al. 2015; Khun 2005; Poncolet et al. 2010; Warner et al. 2009). Yet, a high exposure and sensitivity to hazards does not necessarily lead to an increase in permanent migration. Based on a longitudinal study in Bangladesh, Gray and Mueller (2012: 4) pointed out that "although mobility can serve as a post-disaster coping strategy, it does not do so universally, and disasters in fact can *reduce* mobility by increasing labor needs at origin or by removing the resources necessary to migrate". Even though many flee before a natural hazard to save their lives and others migrate after an environmental event and can indeed cope with its effect, others (have to) remain behind for social, cultural or purely economic reasons. Immobility is then a particular source of their vulnerability. Many families living under conditions of extreme poverty may experience significant barriers to migration – irrespective of the particular hazards – and thereby cannot cope with a disaster or improve their situation in the longer term. They have neither adult male family members who could work as labor migrants, the required resources to facilitate migration, nor the access to the necessary migration networks. These "trapped populations" (Black et al. 2013; Foresight 2011; Poncolet et al. 2010) are forced to adapt to disasters with resources that are their locally available to them. They largely depend on post-disaster aid as well as mutual help and solidarity within the respective community. Yet, as resources are scarce right after a disaster, their recovery is likely to be longer and more difficult. Many enter and stay locked in systems of dependency and indebtedness. The trapped populations, among them many elderly and many female-headed households, are then probably the people who are most vulnerable to natural hazards.

6.4.2 Human Mobility in the Context of More Subtle Environmental Changes

Almost all Bangladeshis are exposed and sensitive to sudden-onset disasters that hit a particular area. Slow-onset processes do, however, reveal a differential kind of vulnerability. Although slow in its onset, *riverbank erosion* is a common threat to people living along the major rivers and on the many Char islands (the riverbed sandbars) and regularly forces people to move their homes to other nearby places or to migrate permanently to cities (Arsenault et al. 2015; Haque et al. 1989; Mutton and Haque 2004; Poncolet et al. 2010). Since 1973, over 158,780 ha of land has been eroded. More than 16,000 people living on the banks of the Jamuna, Ganges, and Padma have been displaced in 2010 alone (IOM 2010).

Coastal erosion is a slow-onset process that intersects with other threats that the people living in Bangladesh's coastal zones face like *salt-water intrusion, storm surges* and *floods*. Even a 50 cm rise in sea-levels by 2050 could displace 26 million from coastal zones of Bangladesh (Biermann et al. 2010; Gemenne 2011). In 2009, at the United Nations General Assembly Bangladesh's Prime Minister Sheikh Hasina thus reiterated the danger posed by *sea-level rise*:

> What is alarming is that a meter rise in sea level would inundate 18 % of our land mass, directly impacting 11 % of our people. […] of the billion people expected to be displaced worldwide by 2050 by climate change factors … one in every seven people in Bangladesh, would be a victim.

Sea-level rise and *salt water intrusion* were first felt by farmers in the South-west of Bangladesh, not in terms of complete loss of livelihoods, but in terms of decreasing (rice) yields that make it more difficult to sustain an agricultural-based life. For agriculturally-based livelihoods, the environmental stress by water-logging and salinization is particularly acute when the soil quality deteriorates rapidly, or when land has to be given up completely. Given the absence of alternative livelihood options in many rural areas, temporary labor migration and/or permanent out-migration are then among the most frequently used adaptive strategies (IOM 2010; Penning-Rowsell et al. 2013; Poncolet et al. 2010; Pouliotte et al. 2009; Rabbani et al. 2013). Yet, the proliferation of the shrimp industry in the South-west and a changing local labor market also drove out many former farmers, while other people were attracted to and actually migrated to this region (Ackerly et al. 2015). Such transformations of local livelihood systems and the broader regional and global political ecology have to be addressed too, when we want to comprehend people's vulnerability and their migration patterns.

In Bangladesh, the impact of *shifting seasons* and *rainfall variability* on local livelihoods and subsequent migration patterns has been studied the least. Food security is an important intermediate variable between climate change, natural hazards or environmental deterioration and people's decision to migrate or flee. As far as the supply of food is concerned, Bangladesh has shown its adaptive capacity to deal with challenges stemming from global food markets, trade agreements or climatic change. There still seems to be untapped potential to produce more food. Yet, most people in rural areas still depend on subsistence production and agricultural day-labor for their food security. A good harvest is based on the availability of water in the right quantity at the right time. If farmers fail to respond to the variability in rainfall by using irrigation, they risk losing (parts of) their production. As poor farmers cannot afford irrigation, too little water during the critical crop-growing period decreases their food production. If no alternative employment opportunities are available locally, labor mobility can become the sole option to secure a household's access to food (Etzold et al. 2014, 2015; Findlay and Geddes 2011; Poncolet et al. 2010). According to Gray and Mueller (2012), a positive significant relationship does exist between crop failures, which are driven by rainfall variability, and long-term migration. However, the propensity to migrate permanently due to crop loss and food insecurity differs strongly among rural households. In case of a drought, landless laborers do not lose their own production, but rather their work. They are more likely to migrate permanently in search of work than members from households who have lost their harvest, but hope to recover at home. People's *sensitivity* to rainfall variability, which is a socio-economic determinant that largely rests on landownership, and not their mere *exposure*, is then the key element to understanding their coping actions and their overall vulnerability to climatic risks. These findings have been verified in the "Where the Rain Falls" study that was carried out in 2011 in northern Bangladesh by UNU-EHS and CARE (see Textbox 6.2).

Textbox 6.2: Rainfall Variability, Food Insecurity and Migration in Northern Bangladesh
The "Where the Rain falls" research in Bangladesh focused on four villages in Kurigram district in Rangpur division in the country's North, which is known for a high incidence of poverty and seasonal food insecurity. The study found that rainfall variability, household food security, and migration patterns are closely intertwined. In the region, agricultural production mostly depends on rainfall, as costly irrigation systems are not widespread. Disruptions of livelihoods, such as natural hazards and significant changes in the annual monsoon cycle, are perceived by the local people as severe livelihood risks, since both have direct effects on food production and incomes. An 'untypical' longer dry period during the monsoon season can, for instance, lead to crop damage and reduced food production. It also contributes to a local increase in food prices and reduces the demand for agricultural laborers throughout the harvest season. As a consequence, small-scale farmers and wage laborers have to reduce their food consumption to cope with the effects of rainfall variability or they will have to seek alternative income sources, for instance through migration.

Half of the interviewed households in the survey had internal migrants in their family. Ninety-seven percent of the migrants were men. Most move-

(continued)

Textbox 6.2 (continued)

ments take place within the country and over short distances. Forty-nine percent of the movements were to cities. Rural-to-rural migration accounted for 47% of all movements. Although the number of out-migrants has increased significantly over the past decades, the vast majority of the people in Kurigram district are not mobile (BBS 2012). Migration is seen by the population as a way to cope during a period of crisis, in particular to avoid or reduce food insecurity (as noted by 79% of respondents), but also as a normal income diversifying activity (as mentioned by 27%). The major reasons to migrate can be ranked as follows: poverty and lack of employment opportunities in home region, then food insecurity, followed by rainfall variability and natural hazards.

The results of the study also show that persisting local patterns of social inequality and food insecurity matter crucially for different groups' propensity to migrate in the context of rainfall variability. The most affluent and food secure people do not need to migrate in order to adapt to the negative effects of rainfall variability, because their livelihoods are already fairly resilient. In stark contrast, the poorest and most food insecure people cannot migrate at all. These most vulnerable 'trapped' households are forced to cope locally with rainfall variability as they neither have adult male family members who could work as labor migrants, nor resources to facilitate migration, nor access to the necessary migration networks.

Some households are 'getting ahead' as migration led to a diversification and translocalization of their livelihood and a reduction of their sensitivity to rainfall variability. Members from other households migrate, because they have been negatively affected by too much or too little rain, and seek to cope with such a temporary crisis. Yet, they are just 'getting by' and can neither get out of poverty, nor reduce their sensitivity to rainfall variability. Some households use migration as an option of "last resort" (Penning-Rowsell et al. 2013) to overcome the worst periods of hunger, but their overall situation and the conditions for those 'left behind' actually deteriorate (Peth and Birtel 2015; Warner and Afifi 2014).

Overall, the case study demonstrated that instead of rainfall variability – as one indicator of climate change – it is social inequality and food insecurity in the region as well as the structural economic differences between more remote rural areas, prospering agricultural regions and the major urban centres that drive migration from the Kurigram region. Nonetheless, labor migration is one of the most important coping strategies of rural households in the context of multiple climatic risks (Ahmed et al. 2012; Etzold et al. 2014, 2015).

To sum up, mobility can be an important way to cope with both a disaster and subtle environmental changes, and their respective long-term effects. Yet, the link between temporary migration after disasters and an increase in permanent migration is neither direct nor clear (Gray and Mueller 2012; Joarder and Miller 2013). Moreover, the long-terms effects of people's post-disaster migrations are not well understood. On the one hand, migration can contribute to enhanced livelihood security, when households have gone translocal in order to diversify their sources of income and to reduce their exposure to environmental, economic and political risks at one particular place. Migration can then be seen as an indicator of social resilience and adequate adaptive capacities to climate change (McLeman and Smit 2006). On the other hand, many migrant households do not benefit from migration as the earned incomes – and thus remittances – remain low, as initial periods of indebtedness cannot be overcome, and as the absence of (male) family members has negative effects on the broader family. In these cases, migration can be seen as an 'erosive coping strategy' that even enhances household's vulnerability and limits their future potentials (Warner and Afifi 2014). Last, not least, the relations between *im*mobility and post-disaster recovery, household vulnerability and long-term societal development, and thus the foregone livelihood opportunities for trapped populations need further investigation (Black et al. 2013).

6.5 Towards Recognizing the Normality of Translocal Lives

In Bangladesh, natural hazards, changing rainfall regimes, sea-level rise, and land degradation certainly play an important role in people's decision to migrate or stay. These environmental drivers should, however, not be rated higher than the normal cycles in rural livelihoods and seasonal patterns of food insecurity, fundamental transformations in the local, national and global political economy, basic social changes and demographic trends, or technological innovations (Black et al. 2011; Foresight 2011). Together, different drivers contribute to the decision of many Bangladeshis to incorporate migration into their livelihood strategies or even force some social groups or individuals to move. Migration patterns are also linked to the specific interplay between places of origin and the destinations. Individuals, families, and communities are rooted in specific places and livelihood systems with quite distinct environmental histories; and they are organized, mobilized, or 'kept in place' through social networks, political structures, and institutions. Over time, multi-faceted migration systems emerge through multi-directional and repeated acts of migration, return, and re-migration, through the 'migration services' of mediating agents, and through the translocal relations of people at places of origin, places of transit, and places of destination. Whether mobility can take place at all, how migration works, and in which direction people move thus often depends on the position of (potential) migrants in existing social networks. Two further examples from our work help to reiterate the multi-dimensionality and normality of migration in Bangladesh.

First, for many inhabitants of the hazard-prone coastal area migration has become an integral part of life. Over the past decades, a significant proportion of the men in coastal areas have become migrant workers in order to sustain their families who continue to live in hazard prone areas. However, despite the ability of individuals to migrate for work, very few choose to move permanently to distant locations. Unless there is no other option, people often prefer to remain close to their places of origin. They frequently move back and forth in order to avoid a drastic rupture with their social network (Ackerly et al. 2015; Mallick 2015).

Second, many people in the Northwest face chronic poverty and seasonal food insecurity, even despite two harvests per year. This so called *Monga* phenomenon is caused by a lack of employment and thus income before the rice harvest. In this context, labor migration is an important coping strategy. At the same time, migration has become a normal livelihood strategy for many households. While some migrants move around and work independently, the trips of others are well organized by intermediaries. A translocal life with rhythmically changing periods of work in other rural and urban areas and at home helps many small-scale farmers and landless laborers to increase their income, diversify their livelihood, handle risks, and enhance their life chances (Etzold et al. 2014; Peth and Birtel 2015).

The case studies indicate that a too linear and static picture of migration as permanent movement of a household from place A to place B should be avoided. Nowadays, many Bangladeshi families organize their livelihoods dynamically across different places – they are living translocal and some even transnational lives in order to earn extra cash-income that is needed for daily consumption or agricultural investments, to overcome regular livelihood crises such as hunger during the annual lean season, to diversify risks and buffer shocks at one place such as a failed harvest due to irregular rainfall, or to invest in their own future. Migrants' everyday life is then characterized through their experience of migration as well as their simultaneous embeddedness in and their social networks across local places (Brickel and Datta 2011; Greiner and Sakdapolrak 2013). In order to understand migration and translocality in and beyond Bangladesh – whether it is in the context of climatic changes or not – we need to consider three aspects in particular: existing migration systems, translocal transfers as well as kinship relations and place-based identities.

6.5.1 Taking Existing Migration Systems into Account

Besides international mobility (Dannecker 2005; Gardner 1995; Rahman 2012; Siddiqui 2005), several labor migration systems co-exist *in* Bangladesh: permanent rural-urban and urban-urban migration, temporary migration to cities, and seasonal labor migration to agricultural regions. Rather than triggering new migration patterns, it is more likely that natural hazards and environmental changes contribute to changes of existing livelihood patterns and migration systems (Findlay and Geddes 2011). The Rainfalls study undertaken in Kurigram District in the North of

Bangladesh, for instance, showed that many people do indeed migrate in the context of rainfall variability and enhanced food security. Their labor mobility, however, takes place within established migration systems and within existing social networks. People's access to migration opportunities and their choice of destinations reflects existing patterns of social inequality. Members from more affluent households are more likely to move to urban destinations for secure employment in the formal economy or for higher education. The rural 'middle class' either goes to cities like Dhaka to work in the garments industries, the construction sector, or the informal economy. Such movements are normally self-organized or assisted by family members, or temporary moves to other rural destinations throughout the harvest seasons. Agricultural migration is facilitated by middlemen who negotiate wages with employers and organize transport, accommodation, and food for a group of labor migrants. But not everybody can benefit from such migration systems. The poorest people in rural communities often cannot afford the initial investments needed for migration, nor do they have access to necessary networks or even sometimes, the physical capability to migrate at all. These households remain trapped in Bangladesh's rural spaces of vulnerability (Etzold et al. 2014; Poncolet et al. 2010).

6.5.2 Investigate Translocal Flows and Their Effects

Interactions and transfers within translocal networks are highly important for migrant workers' families, the respective communities, and the economies of migrant-sending- and money-receiving nations. Officially recorded remittances to Bangladesh have steadily increased over the past 40 years. In 2013, 'overseas workers' from Bangladesh sent home 13.9 Billion US$, which accounts for more than 9 % of its GDP.[1] There is no official data on the transactions of internal migrants, but the Rainfalls study in Kurigram, one of the poorest districts in Bangladesh, revealed that the migrants' contribution to households' income is viewed as 'substantial' by more than half of the 150 survey respondents (Ahmed et al. 2012). Sending money has become easier, quicker, cheaper, and safer with the introduction of e-remitting and e-banking via mobile phones (Sterly 2015). Remittances are spent on migrant families' food consumption, debt repayment, health care and education, and business or agricultural investments (Ahmed et al. 2012; Peth and Birtel 2015). Migrants do not only bring home money and consumer products, but also new knowledge, ideas, values, and norms. Material and 'social remittances' play important roles in the everyday lives and identity formations of translocal families and contribute crucially to economic, social and political transformations in Bangladesh (Dannecker 2005; Gardner 1995, 2009; Rahman and Fee 2012). Both types of transfers also leave their marks in space as remitted money is invested back home to buy agricultural land or to build a house, or to buy mobile phones as lifestyles change due to newly introduced products.

---

[1] Since 1976, the total sum of remittances to Bangladesh steadily increased from 19 Million US$ in 1976 to the maximum of 14.2 Billion US$ in 2012 (World Bank n.d).

6.5.3 Acknowledge Peoples Identities and Rootedness in Places

Bangladeshis generally "have a strong sense of home and rootedness" (Gardner 2009: 233), which is maintained through kin connections and reaffirmed through place-based identities. In a study of the "culture of slum dwellers," Kumar Das (2003) questioned a too-simplistic understanding of migration in terms of the most prevalent pull and push factors. He argues that networks that have developed between places of origin and the city, kinship ties and relations to political leaders do structure people's migration trajectories. 'Translocal social capital' can, in turn, become an important asset in times of crises as research on the coping strategies of Dhaka's slum dwellers during the 2007/2008 food price hike has shown (Zingel et al. 2011). Translocal relations are carefully maintained through transfers and through regular mobile phone calls. The home village is visited regularly, in particular, for traditional festivities. Even if people came to Dhaka decades ago, frequent reference to one's home is being made. Strong emotional bonds are not only reaffirmed among relatives, but also among people from the same village or district, who often live in the same neighborhood and support one another, for instance, with finding a job or getting access to public services (Kumar Das 2003). Maintaining close relations to people at multiple places – people with similar origin in the urban neighborhood, other people in the home community and family members in the city, back home or at other place in the world – is a crucial aspect of translocal life. Specific socio-spatial constructions like the *desh*, the 'homeland' of transnational migrants, or the *bari*, the rural 'home village' of the family father, have thus not lost importance for Bangladeshi migrants' sense of belonging (Etzold 2016; Gardner 2009; Kuhn 2003; Peth and Birtel 2015).

We argue that most research on climate change adaptation and human mobility in Bangladesh disregards the already existing normality of mobility for rural communities, the flows within translocal households that are simultaneously embedded in multiple places, the everyday experience of translocal lives, and the significance of place-based identities. In the political realm, a too simplistic, too static and culturally uniformed understanding of migration and mobility does not only misrepresent the social reality. It also leads to the wrong solutions.

6.6 De-constructing the Politics of Climate Change Adaptation

6.6.1 Playing the Numbers Game in Bangladesh

Narratives of displacement and migration play a significant role in political discourses of climate change adaptation. Numbers matter crucially in these discourses. Myers (2002) has prominently argued that 25 million people were displaced worldwide in 1995 due to environmental disruptions; by 2050, 200 million people would be displaced. Assuming a temperature increase of 4 °C by the end of twenty-first century, Nicholls et al. (2011) state that up to 187 million people could be forcibly displaced worldwide by sea-level rise and associated consequences. For Bangladesh, Biermann et al. (2010) projected that 26 million people would be affected and displaced by storm surges and sea-level rise by the year 2050. This is close to the estimate of 29 million people (one-seventh of Bangladesh's projected population of 202 million people for 2050) in the aforementioned quote by Bangladesh's Prime Minister. Ahmed and Neelormi (2008) estimated that 250,000 people might be displaced each year due to climate hazards under a moderate climate change scenario.

It is likely that the number of people who (temporarily) move from Bangladesh's rural areas to other places will increase further in the light of climate change, environmental degradation, *and* economic globalization. Yet, we should be very careful with such number games. First, all too often the blunt estimates simply lack an empirical basis and the methods of estimation or projection are neither displayed nor discussed. Vague estimates are then circulated and thereby reinforced, while the original data sources and remarks on the validity of the estimates somehow get lost. What is left is the blank number, which might be manipulated for political purposes (Bettini 2013; Irfanullah 2013). Second, the exact reasons why people are displaced – or do they move voluntarily? – are often not considered adequately in the estimates. Gemenne (2011) argued that the existing estimates of the number of people likely to be displaced from flood-prone areas in Bangladesh do not adequately distinguish between the different environmental risks (e.g. flooding, cyclones, sea-level rise) that people are exposed to. They also do not consider other crucial drivers of migration or immobility, such as poverty levels, local economic transformations, political conflicts or alternative adaptation strategies. Third, the underlying assumptions of most estimates and projections are overly simplistic. People are displaced by nature for good: they leave once and for all, they do not come back, they do not move forward. Migration is thereby portrayed as a singular and linear process. This is crudely environmentally deterministic and reductionist as all other social, cultural, economic, political and spatial drivers that contribute to migration decisions are simply not considered (Black et al. 2011). It also denies people their capacity to anticipate changes in their environment, to cope with shocks to their livelihoods, and to adapt to more structural transformations in the area where they live. In short, there is no place for human agency in such projections. Fourth, many projections use the numbers of thousands and millions of 'climate refugees' in a Neo-Malthusian 'alarmist scenario' to demonstrate that something is going wrong. This might be legitimate, if one seeks to raise attention for climate change mitigation and adaptation. In the policy discourse, however, human mobility is thereby framed as something 'bad' or a 'failure of climate change adaptation' that needs to be avoided, or at least 'better managed' (Bettini 2013; Mayer et al. 2013; Saul 2011). Criticizing such negative connotations, we argue that human mobility is a right in itself. Migration and translocality can be seen as adequate ways to live with to environmental, economic and political risks at multiple places.

6.6.2 Political Responses to 'Environmentally-Induced Migration': Two Perspectives

In Bangladesh, quite a number of policy documents and strategy papers make specific reference to population displacement as a consequence of environmental disruptions and climate change. The National Adaptation Plan of Action (NAPA) of 2005 and the Bangladesh Climate Change Strategy and Action Plan (BCCSAP) of 2009 have been designed in order to enhance the nation's adaptive capacities and increase its resilience in the context of climate change. As indicated, one of the principle challenges in the political debate about climate change and migration lies in the way how mobility is discursively framed. It makes a big difference whether migration in the context of climate change is seen as a 'failure of adaptation' and thereby the migrants as a 'security risk' or as an 'adequate way of adaptation' and the migrants as 'agents of change' (Martin et al. 2013; Mayer et al. 2013; Naser 2014). The consequence of the first notion would be to adjust and develop policy tools and development interventions in order to strengthen people's capacities to mitigate risks and thereby reduce their vulnerability to natural hazards and environmental degradation so that they can 'stay at home.' One example involves common assumptions about sea-level rise and people's protection in coastal areas. Although the rate of sea-level rise is rather slow and the long-term effects in the Ganges-Brahmaputra-Delta are not clear, the current discussions in Bangladesh concentrates on losses of habitat and livelihoods and on projections of the number of people who might be displaced. The straightforward political solution points to early warning systems, emergency aid, and infrastructure projects such as flood and cyclone shelters, in which people living in highly exposed areas can temporarily seek protection. Projects that support people to maintain their livelihoods, develop new livelihood opportunities and cope with cyclones and floods – maybe with the help of micro-credits and/or micro-insurance – without having to flee or temporarily move are also discussed (Arsenault et al. 2015; Mallick 2015). Besides people's protection and prevention of loss and damage, the rational then is to 'keep people in their place.' The fears of the Indian public and policy-makers of mass migration from Bangladesh might follow from such a perception of migration as 'failure of adaptation' to climate change. Besides keeping people in their place, better migration management and enhanced border security are then logical political responses by India (Panda 2015). While it is clear that bilateral regional and international cooperation is necessary in this field, a closure of borders and other more rigid measures against migrants would only lead to further suffering and human rights violations at the Indian-Bangladeshi border. And it would completely ignore the reality of everyday life – thousands of Bangladeshis already live a transnational life and cross the border frequently, if not daily.

In contrast to searching for 'local' solutions to climate change adaptation and to conceive of environmentally-induced migration as a problem in itself, the political consequence of seeing 'migration as an adequate way of adaptation' would be to embrace mobility. In this sense, planned relocation, voluntary resettlement, temporary or permanent labor migration – both nationally and internationally – as well as translocal lifestyles and transnational diaspora relations could be actively fostered in order to enhance the Bangladeshi people's resilience in times of climate change. The appropriate policies would then look very different from those sketched above. They could center on an improved management of migration and institutional support for migrant workers, and thus on fundamental aspects of social protection, workers' rights, and human rights at the multiple destinations of migrants. For Bangladeshis, these destinations are mainly Dhaka, Chittagong, and Khulna, where thousands of rickshaw pullers, garments factory workers, and day-laborers seek to sustain their lives with 'dirty, dangerous and demanding' work; the prospering rural areas, where seasonal agricultural workers often labor under harsh conditions; and in particular foreign destinations such as Dubai, Qatar, and Kuala Lumpur, where tens of thousands Bangladeshis work in the construction sector, in kitchens and as cleaners – too often under exploitative conditions. The way migration is organized – whether it is through international manpower agencies or *Sadars* who facilitate temporary labor migration within the nation – is crucial for labor conditions and the benefits that migrants can derive from their labor. Other political fora than those of the international climate regime are relevant as a basis for improving migrant workers' labor conditions, for instance, the "International Convention on the Protection of the Rights of All Migrant Workers" or the "ILO Declaration on Fundamental Principles and Rights at Work".

Another political option is to support everyday translocal life. Technological advances like the internet and mobile phone communication have brought more significant positive changes for translocal families than any measures that have targeted migrants as such (Peth and Birtel 2015; Sterly 2015). The Bangladeshi state can further support translocal households' self-organizing capacity by investing in its information and communication technology (ICT) infrastructure and by reducing barriers to remittances. Without a doubt the transfers of migrant laborers have transformed the rural areas. They contribute to diversifying migrant families' livelihoods and reducing the vulnerability of those who continue to live in rural places. Every other disaster that happens in Bangladesh proves the significance of migrants' remittances for survival and food security during disaster and for recovery and enhanced resilience thereafter. But not only financial transfers are needed. Social remittances might be even more relevant in the long run. If female members of a household do migrate, they might come back later with more self-confidence and skills that they have acquired under quite different social, cultural and economic conditions (Dannecker 2005). The experience and knowledge of people who have been displaced or who have moved voluntarily, but return, is a potential source for families and communities to better manage the new risks arising in the wake of climate change, environmental degradation, social change and economic transformation.

6.7 Conclusion

Without a doubt, natural hazards, subtle environmental changes and weather-related risks nowadays play a decisive role in shaping the vulnerability of Bangladeshi people, and they do certainly influence existing migration patterns and often even lead to the permanent displacement of people from their homes. Yet, we should not believe too simplistic – and possibly naturally deterministic – accounts of 'climate refugees' and 'environmental migrants', who have simply fled natural hazards and deteriorating environmental conditions. Complex migration processes cannot be determined by nature. They are rather structured by people's perceptions of environmental, economic and political changes, by their everyday experience, their social and cultural embeddedness, and by their (in)ability to see and take on livelihood opportunities at multiple places.

In this chapter, we have moreover argued that 'environmentally-induced migration' i.e. migration patterns that are linked to but not solely caused by climate change, natural hazards or environmental degradation, should generally not be regarded by policy-makers and the public as a local failure of adaption to climate change. Migration might rather be seen as a normal part of life in an increasingly interconnected world and as an adequate strategy of people who seek to diversify and translocalize their livelihoods. Translocal households might be better able to respond to shocks at one place and thereby also more resilient in times of climate change (Sakdapolrak 2014). Yet, too many Bangladeshi people have migrated and now work as agricultural laborers in rural areas or as day-laborers in cities under disadvantageous and even exploitative labor conditions, which do *not* allow them and their families to even meet their basic needs and to enhance their future life chances. Instead, many migrant workers are drawn into a vicious cycle of poverty and indebtedness as migration is often costly in itself, as they cannot cover their initial investments, and as they cannot meet their families' expectations. Nonetheless, those households might be even more vulnerable, who cannot use migration at all as an adaptation strategy in the context of risks or as an option to improve their lives; those families that are 'trapped' in a deteriorating environment, in poverty, and in a disadvantageous social position. In general, we thus plead for more political support for and enhanced social protection of all vulnerable groups, and in particular for migrants, relocated persons, displaced people and refugees, and their translocal families.

References

Ackerly, B., Anam, M., & Giligan, J. (2015). Environment, political economies and livelihood change. In B. Mallick & B. Etzold (Eds.), *Environment, migration and adaptation. Evidence and politics of climate change in Bangladesh* (Vol. 1, pp. 27–39). Dhaka: AHDPH.

Adger, N. W. (2006). Vulnerability. *Global Environmental Change, 16*(3), 268–281.

Afifi, T., & Jäger, J. (Eds.). (2010). *Environment, forced migration and social vulnerability*. Berlin: Springer.

Afsar, R. (2005). Internal migration and the development nexus: The case of Bangladesh. In T. Siddiqui (Ed.), *Migration and development. Pro-poor policy choices* (pp. 39–69). Dhaka: The University Press Limited.

Ahmed, A. U., & Neelormi, S. (2008). *Loss of livelihoods and forced displacements in bangladesh: Whither facilitated international migration?* Dhaka: Campaign for Sustainable Rural Livelihoods (CSRL).

Ahmed, A. U. A., Hassan, S. R., Etzold, B., & Neerlormi, S. (2012). *Rainfall, food security and human mobility. Case study: Bangladesh. Results from Kurigram district, Rangpur* ("Where the Rain Falls" Project, Report No. 2). Bonn: United Nations University Institute for Environment and Human Security (UNU-EHS).

Akter, S., & Fatema, N. (2015). Coping with flood risks: Microcredit versus micro-insurance. In B. Mallick & B. Etzold (Eds.), *Environment, migration and adaptation – Evidence and politics of climate change in Bangladesh* (pp. 175–191). Dhaka: AHDPH.

Arsenault, M.-P., Azam, M. N., & Ahmad, S. (2015). Riverbank erosion and migration. In B. Mallick & B. Etzold (Eds.), *Environment, migration and adaptation. Evidence and politics of climate change in Bangladesh* (Vol. 1, pp. 41–61). Dhaka: AHDPH.

Bangladesh Bureau of Statistics (BBS). (2012). *Population census 2011. Socio-economic and demographic report*. Dhaka: Bangladesh Bureau of Statistics, BBS.

Bangladesh Garment Manufacturers and Exporters Association (BGMEA). (n.d.). Trade Information http://www.bgmea.com.bd/home/pages/TradeInformation#.Uo2-I-Ly-no Accessed 1 Mar 2016.

Bettini, G. (2013). Climate Barbarians at the gate? A critique of apocalyptic narratives on 'climate refugees'. *Geoforum, 45*(3), 63–72.

Biermann, F., & Boas, I. (2010). Preparing for a warmer world: Towards a global governance system to protect climate refugees. *Global Environmental Politics, 10*(1), 60–88.

Black, R., Adger, N., Arnell, N. A., Dercon, S., Geddes, A., & Thomas, D. S. G. (2011). The effect of environmental change on human migration. *Global Environmental Change, 21S*, S3–S11. doi:10.1016/j.gloenvcha.2011.10.001.

Black, R., Arnell, N. W., Adger, W. N., Thomas, D., & Geddes, A. (2013). Migration, immobility and displacement outcomes following extreme events. *Environmental Science & Policy, 27*, S32–S43.

Bohle, H.-G. (2007). *Living with vulnerability. Livelihoods and human security* (InterSecTions No.6/2007). Bonn: United Nations University - Institute for Environment and Human Security, UNU-EHS.

Brickel, K., & Datta, A. (2011). Introduction: Translocal geographies. In K. Brickel & A. Datta (Eds.), *Translocal geographies. Spaces, places, connections* (pp. 3–20). London: Ashgate.

Dannecker, P. (2002). *Between conformity and resistance. Women garment workers in Bangladesh*. Dhaka: The University Press Limited.

Dannecker, P. (2005). Transnational migration and the transformation of gender relations: The case of Bangladeshi labour migrants. *Current Sociology, 53*(4), 655–674. doi:10.1177/0011392105052720.

Etzold, B. (2016). Migration, informal labour and (trans)local productions of urban space – The case of Dhaka's street food vendors. *Population, Space and Place, 22*(2), 170–184. doi:10.1002/psp.1893.

Etzold, B., Ahmed, A. U., Hassan, S. R., & Neelormi, S. (2014). Clouds gather in the sky, but no rain falls. Vulnerability to rainfall variability and food insecurity in Northern Bangladesh and its effects on migration. *Climate and Development, 6*(1), 18–27. doi:10.1080/17565529.2013.833078.

Etzold, B., Ahmed, A. U., Hassan, S. R., & Afiffi, T. (2015). Climate change, hunger or social inequality – Which one drives migration? In B. Mallick & B. Etzold (Eds.), *Environment, migration and adaptation. Evidence and politics of climate change in Bangladesh* (Vol. 1, pp. 79–98). Dhaka: AHDPH.

Findlay, A., & Geddes, A. (2011). Critical views on the relationship between climate change and migration: Some insights from the experience in Bangladesh. In E. Piguet, A. Pécoud, & P. D. Guchteneire (Eds.), *Migration and climate change* (pp. 138–159). Paris: UNESCO, Cambridge University Press.

Foresight. (2011). *Migration and global environmental change – Final project report*. London: UK Government Office for Science.

Füssel, H.-M. (2007). Vulnerability: A generally applicable conceptual framework for climate change research. *Global Environmental Change, 17*(2), 155–167.

Gardner, K. (1995). *Global migrants, local lives: Travel and transformation in rural Bangladesh: Travel and transformation in rural Bangladesh*. Oxford: Oxford University Press.

Gardner, K. (2009). Lives in motion: The life-course, movement and migration in Bangladesh. *Journal of South Asian Development, 4*(2), 229–251. doi:10.1177/097317410900400204.

Gehlich-Shillabeer, M. (2008). Poverty alleviation or poverty traps? Microcredits and vulnerability in Bangladesh. *Disaster Prevention and Management, 17*(3), 396–409.

Gemenne, F. (2011). Climate-induced population displacements in a 4°C+ world. *Philosophical Transactions of the Royal Society A: Mathematical, Physical and Engineering Sciences, 369*, 182–195. doi:10.1098/rsta.2010.0287.

Gray, C. L., & Mueller, V. (2012). Natural disasters and population mobility in Bangladesh. *PNAS, 109*(16), 6000–6005. doi:10.1073/pnas.1115944109.

Greiner, C., & Sakdapolrak, P. (2013). Translocality: Concepts, applications and emerging research perspectives. *Geography Compass, 7*(5), 373–384. doi:10.1111/gec3.12048.

Haan, A., Brock, K., Carswell, G., Coulibaly, N., Seba, H., & Toufique, K. A. (2000). *Migration and livelihoods: Case studies in Bangladesh, Ethiopia and Mali*. Brighton: Institute of Development Studies, IDS.

Haque, C. E., & Zaman, M. Q. (1989). Coping with riverbank erosion hazard and displacement in Bangladesh: Survival strategies and adjustments. *Disasters, 13*(4), 300–314.

Intergovernmental Panel on Climate Change (IPCC). (2007). *Climate change 2007: Impacts, adaptation and vulnerability. Contribution of Working Group II to the fourth assessment report of the Intergovernmental Panel on Climate Change*. Cambridge: Cambridge University Press.

Intergovernmental Panel on Climate Change (IPCC). (2014a). *Climate change 2014: Impacts, adaptation and vulnerability. Part B: Regional aspects. Working Group II contribution to the fifth assessment report of the Intergovernmental Panel on Climate Change*. Cambridge: Cambridge University Press.

Intergovernmental Panel on Climate Change (IPCC). (2014b). *Climate change 2014: Impacts, adaptation and vulnerability. Summary for policy makers. Working Group II contribution to the fifth assessment report of the Intergovernmental Panel on Climate Change*. Cambridge: Cambridge University Press.

International Organization for Migration (IOM). (2010). *Assessing the evidence: Environment, climate change and migration in Bangladesh*. Dhaka: International Organization for Migration (IOM).

Irfanullah, H. M. (2013). In search of '35 million climate refugees' from Bangladesh. *International Journal of Environmental Studies, 70*(6), 841–843.

Islam, I., Akther, S., Jahan, N., & Hossain, M. I. (2015). Displacement and migration from cyclone-affected coastal areas. In B. Mallick & B. Etzold (Eds.), *Environment, migration and adaptation – Evidence and politics of climate change in Bangladesh* (pp. 141–160). Dhaka: AHDPH.

Joarder, M. A. M., & Miller, P. W. (2013). Factors affecting whether environmental migration is temporary or permanent: Evidence from Bangladesh. *Global Environmental Change-Human and Policy Dimensions, 23*(6), 1511–1524. doi:10.1016/j.gloenvcha.2013.07.026.

Kartiki, K. (2011). Climate change and migration: A case study from rural Bangladesh. *Gender and Development, 19*(1), 23–38.

Kuhn, R. (2003). Identities in motion: Social exchange networks and rural-urban migration in Bangladesh. *Contributions to Indian Sociology, 37*(1–2), 311–337. doi:10.1177/006996670303700113.

Khun, R. (2005). *The determinants of family and individual migration: A case-study of rural Bangladesh* (Working Paper). Boulder: University of Colorado, IBS Population Program.

Kumar Das, T. (2003). *Culture of slum dwellers: A study of a slum in Dhaka*. Dhaka: Boipatro.

Levitt, P., & Glick Schiller, N. (2004). Conceptualizing simultaneity: A transnational social field perspective on society. *International Migration Review, 38*(3), 1002–1039.

Mallick, B. (2015). Cyclone-induced migration in southwest coastal Bangladesh. In B. Mallick & B. Etzold (Eds.), *Environment, migration and adaptation. Evidence and politics of climate change in Bangladesh* (pp. 119–140). Dhaka: AHDPH.

Mallick, B., & Etzold, B. (Eds.). (2015). *Environment, migration and adaptation. Evidence and politics of climate change in Bangladesh*. Dhaka: AHDPH.

Mallick, B., & Siddiqui, T. (2015). Disaster-induced migration and adaptation discourse in Bangladesh. In F. Hillmann, M. Pahl, B. Rafflenbeul, & H. Sterly (Eds.), *Migration, environmental change and adaptation in a regional perspective* (pp. 164–185). London: Palgrave Macmillan.

Mallick, B., & Vogt, J. (2012). Cyclone, coastal society and migration: Empirical evidence from Bangladesh. *International Development Planning Review, 34*(3), 217–240.

Mallick, B., & Vogt, J. (2014). Population displacement after cyclone and its consequences: Empirical evidence from coastal Bangladesh. *Natural Hazards, 73*(3), 191–212.

Martin, M., Billah, M., Siddiqui, T., Black, R., & Kniveton, D. (2013). *Policy analysis: Climate change and migration in Bangladesh* (Working Paper 2). Refugee and Migratory Movements Research Unit (RMMRU), Sussex Centre for Migration Research (SCMR).

Mayer, B., Boas, I., Ewing, J. J., Baillat, A., & Das, U. K. (2013). Governing environmentally-related migration in Bangladesh: Responsibilities, security and the causality problem. *Asian and Pacific Migration Journal, 22*(2), 177–198.

McLeman, R., & Smit, B. (2006). Migration as an adaptation to climate change. *Climatic Change, 76*(1–2), 31–53. doi:10.1007/00584-005-900.

Mutton, D., & Haque, C. E. (2004). Human vulnerability, dislocation and resettlement: Adaptation processes of river-bank erosion-induced displaces in Bangladesh. *Disasters, 28*(1), 41–62.

Myers, N. (2002). Environmental refugees: A growing phenomenon of the 21st century. *Philosophic Transactions: Biological Sciences, 357*(1420), 609–613.

Naser, M. M. (2014). Climate change and migration: Law and policy perspectives in Bangladesh. *Asian Journal of Law and Society, FirstView*, 1–19. doi:10.1017/als.2014.7.

Nicholls, R., Marinova, N., Lowe, J., Brown, S., Vellinga, P., de Gusmao, D., et al. (2011). Sea-level rise and its possible impacts given a 'beyond 4°C world' in the twenty-first century. *Philosophical Transactions – Royal Society. Mathematical, Physical and Engineering Sciences, 369*, 161–181.

Panda, A. (2015). Climate-induced migration and interdependent vulnerabilities between Bangladesh and India. In B. Mallick & B. Etzold (Eds.), *Environment, migration and adaptation. Evidence and politics of climate change in Bangladesh* (pp. 195–211). Dhaka: AHDPH.

Paul, B. K. (2005). Evidence against disaster-induced migration: The 2004 tornado in north-central Bangladesh. *Disasters, 29*(4), 370–385.

Penning-Rowsell, E. C., Sultana, P., & Thompson, P. M. (2013). The 'last resort'? Population movement in response to climate-related hazards in Bangladesh. *Environmental Science & Policy, 27*, S44–S59. doi:10.1016/j.envsci.2012.03.009.

Peth, S., & Birtel, S. (2015). Translocal livelihoods and labor migration in Bangladesh: Migration decisions in the context of multiple insecurities and a changing environment. In B. Mallick & B. Etzold (Eds.), *Environment, migration and adaptation – Evidence and politics of climate change in Bangladesh* (pp. 99–118). Dhaka: AHDPH.

Piguet, E., Pécoud, A., & Guchteneire, P. d. (Eds.). (2011a). *Migration and climate change*. Paris: UNESCO, Cambridge University Press.

Piguet, E., Pécoud, A., & Guchteneire, P. d. (2011b). Migration and climate change: An overview. *Refugee Survey Quarterly, 30*(3), 1–23. doi:10.1093/rsq/hdr006.

Poncolet, A., Gemenne, F., Martinello, M., & Bousetta, H. (2010). A country made for disasters: Environmental vulnerability and forced migration in Bangladesh. In T. Afifi & J. Jäger (Eds.), *Environment, forced migration and social vulnerability* (pp. 211–222). Berlin: Springer.

Pouliotte, J., Smit, B., & Westerhoff, L. (2009). Adaptation and development: Livelihoods and climate change in Subarnabad, Bangladesh. *Climate and Development, 1*(1), 31–46.

Rabbani, G., Rahman, A., & Mainuddin, K. (2013). Salinity-induced loss and damage to farming households in coastal Bangladesh. *International Journal of Global Warming, 5*(4), 400–415.

Rahman, M. M. (2012). Bangladeshi labour migration to the Gulf states: Patterns of recruitment and processes. *Canadian Journal of Development Studies, 33*(2), 214–230. doi:http://dx.doi.org/10.1080/02255189.2012.689612.

Rahman, M. M., & Fee, L. K. (2012). Towards a sociology of migrant remittances in Asia: Conceptual and methodological challenges. *Journal of Ethnic and Migration Studies, 38*(4), 689–706. doi:http://dx.doi.org/10.1080/1369183X.2012.659129.

Sakdapolrak, P. (2014). *Building resilience through translocality. Climate change, migration and social resilience of rural communities in Thailand* (TransRe Working Paper, No. 1). Bonn: Department of Geography, University of Bonn.

Salway, S., Rahman, S., & Jesmin, S. (2003). A profile of women's work participation among the urban poor of Dhaka. *World Development, 31*(5), 881–901.

Samers, M. (2010). *Migration* (Key ideas in geography). London: Routledge.

Saul, B. (2011). The security risks of climate change displacement in Bangladesh. *The Journal of Human Security, 8*(2), 5–35.

Siddiqui, T. (2005). International migration as a livelihood strategy of the poor: The Bangladesh case. In T. Siddiqui (Ed.), *Migration and development. Pro-poor policy choices* (pp. 71–107). Dhaka: The University Press Limited.

Smit, B., & Wandel, J. (2006). Adaptation, adaptive capacity and vulnerability. *Global Environmental Change, 16*(3), 282–292.

Sterly, H. (2015). "Without mobile suppose I had to go there" – Mobile communication in translocal family constellations in Bangladesh. *Asien – The German Journal on Contemporary Asia, 131*, 31–46.

Thieme, S. (2007). Sustaining livelihoods in multi-local settings: Possible theoretical linkages between transnational migration and livelihood studies. *Mobilities, 3*(1), 51–71. doi:http://dx.doi.org/10.1080/17450100701797315.

United Nations Department of Economic and Social Affairs, Population Division (UN/DESA). (2014). *World urbanization prospects: The 2014 revision. highlights*. New York: United Nations Department of Economic and Social Affairs, Population Division.

United Nations Development Programme (UNDP). (2014). *Human development report. Sustaining human progress: Reducing vulnerabilities and building resilience*. New York: United Nations Development Programme.

van Schendel, W. (2009). *A history of Bangladesh*. New York: Cambridge University Press.

Warner, K., & Afifi, T. (2014). Where the rain falls: Evidence from 8 countries on how vulnerable households use migration to manage the risk of rainfall variability and food insecurity. *Climate & Development, 6*(1), 1–17. doi:10.1080/17565529.2013.835707.

Warner, K., Ehrhart, C., de Sherbin, A., Adano, S., & Chai-Onn, T. (2009). *In search of shelter. Mapping the effects of climate change on human migration and displacement*: United Nations University – Institute for Environment and Human Security (UNU-EHS); CARE International; CIESIN Columbia University; UNHCR; The World Bank.

Warner, K., Hamza, M., Oliver-Smith, A., Renaud, F., & Julca, A. (2010). Climate change, environmental degradation and migration. *Natural Hazards, 55*(3), 689–715. doi:10.1007/s11069-009-9419-7.

Wisner, B., Blaikie, P., Cannon, T., & Davis, I. (2004). *At risk – Natural hazards, people's vulnerability, and disasters* (2nd ed.). London: Routledge.

World Bank. (2010). *Economics of adaptation to climate change*. Washington, DC: The World Bank Group.

World Bank. (n.d). Personal remittances, received (current US$) Open Data. http://data.worldbank.org/indicator/BX.TRF.PWKR.CD.DT?page=1. Accessed 1 Mar 2016.

Zingel, W.-P., Keck, M., Etzold, B., & Bohle, H.-G. (2011). Urban food security and health status of the poor in Dhaka, Bangladesh. In A. Krämer, M. Khan, & F. Kraas (Eds.), *Health in megacities and urban areas* (Vol. 11, pp. 301–319). Heidelberg: Springer.

Part III
Small Islands

Chapter 7
Good Fishing in Rising Seas: Kandholhudhoo, Dhuvaafaru, and the Need for a Development-Based Migration Policy in the Maldives

Andrea C. Simonelli

7.1 Introduction

The long-term processes of climate change have already become apparent. All across the globe, variation in seasonal weather patterns is affecting those who work in environmentally-based livelihoods. No longer can people count on knowing exactly when the rain will fall, when the rains will stop, or when temperatures will be favorable. These uncertainties hit hard, especially for those whose subsistence fully depends on this knowledge. As such, many have to resort to alternative methods to support themselves, including finding other sources of income. To do this, many households choose a variant of migration as a strategy for making ends meet. Some will send a family member to a nearby city temporarily, others may migrate with their entire households for a season. In essence, many households are finding it necessary to include additional labor activities into their main livelihoods as a way to adapt to climate stresses. In cases such as these, migration can become a strategy of risk management, so that a household is able to compete against forces beyond their control, in order to sustain their families.

Migration employed as adaptation is beginning to be seen as an active strategy of individual/household agency. In a world of new uncertainties, migration, in its many forms, can be seen an option for household empowerment. But not every household whose livelihood is based on the environment can employ this strategy. Workers in agriculture and livestock have certain advantages in operationalizing migration strategies as compared to those in other environmentally-based livelihoods. Living on contiguous land, people can use basic ground transportation to migrate such as a car, bus, train, or in the worst case scenario- walking. For fishing communities in

---

A.C. Simonelli (✉)
Adaptation Strategies International, University of Louisville, Louisville, KY, USA
e-mail: acsimonelli@gmail.com

© Springer International Publishing Switzerland 2016
A. Milan et al. (eds.), *Migration, Risk Management and Climate Change: Evidence and Policy Responses*, Global Migration Issues 6,
DOI 10.1007/978-3-319-42922-9_7

isolated island locales, migration may also be necessary, but cannot be realized without a boat or sea plane. Literature on migration as adaptation to climate change tends to stem from livelihood stresses and the ways in which households and communities need to alter the main means of subsistence and income (Argawal and Perrin 2008; Warner et al. 2009; Warner and Afifi 2014). Conversely, there are people for whom livelihoods are not the concern, but the structural integrity of their place of habitation. What does one do when livelihoods are solid but one's island is not? There will be some cases in which migration as adaptation is the only option; many small atolls close to sea level will eventually be lost to sea-level rise and thus their entire populations will need to relocate. Descriptions of this process then include the issue of sea water leaching into ground water and poisoning any possible subsistence agriculture. While this may be the case, some particular livelihood activities are stable enough to allow some populations to endure these particular processes of degradation. But doing so does not alleviate the need for eventual migration, instead, it changes its form and its functional operationalization.

The Maldives shares common vulnerabilities to other small island developing states (SIDS), but also some important differences. Its significant income through tourist dollars has helped its status within the international community (Athukorala 2014) as well as lift the incomes of many families (Simonelli 2014a). However, not all Maldivians live close enough to a resort to have access to this type of employment. Those who still depend on the environment for their primary occupation are vulnerable to the same climatic stressors as those on many other islands. These include the effects of rising sea surface temperatures, ocean acidification, larger and more intense storms, and sea level rise. While many islands face similar climate stressors, not all islands bear them in the same way. Small tropical islands are particularly susceptible to atmospheric and oceanic circulation, which is especially apparent during El Niño phases. Additionally, even a modest rise in sea levels is likely to result in significant erosion and submersion of land, increased flood hazard, salinization of freshwater aquafers, and the loss of protective coral reefs and sand beaches increasing exposure to hurricanes and storm surges (Pelling and Uitto 2001). Sea level rise and increased water temperatures are projected to accelerate beach and coastal erosion and cause degradation of natural coastal defenses including mangrove and coral reefs. This would impact negatively on water supply, fisheries productivity and the tourism industries in small islands (Hare et al. 2011). In the Pacific, sea level rise is expected to further impact islands through extreme rainfall and storms, which will cause erosion and run-off from terrestrial catchments, further degrading inshore seagrass and coral habitats (Butler et al. 2014). These climate impacts seriously threaten islands, but small islands in particular. Turvey (2007) argues that small islands are more vulnerable to these impacts than larger island countries based on four dimensions of geographic vulnerability. Considering the Maldives in this analysis, small islands are more vulnerable due to inundation risk, their remote nature, vulnerability to urbanization and overpopulation pressures, and vulnerability to natural disasters.

Like the small islands of the Pacific, the Maldives will face large scale deterioration and while they are small islands, they are also atoll islands, which have their own distinct challenges as compared to small islands on a purely spatial scale. Barnett and Adger (2003) explain,

> *Atolls are rings of coral reefs that enclose a lagoon... Atolls have common environmental problems that render them particularly vulnerable to climate change. They generally have very high population densities, meaning that large numbers of people are potentially exposed to single events... Water reserves on atolls are restricted to a narrow subterranean freshwater lens easily contaminated by salt water and human and industrial wastes (UNEP 1999). These freshwater lenses become depleted in times of low rainfall. (:322).*

Atolls are low lying and dynamic, accumulations of sands and gravels which rely on locally generated sediments making them particularly vulnerable to sea level rise (Kench et al. 2005). Being small, isolated, geographically low-lying and with poor-quality soils, atolls face not only physical challenges based on climate effects, but also social and economic vulnerabilities that will ultimately affect the livelihoods of atoll island inhabitants.

7.2 Climate Related Livelihood Vulnerability

Livelihood vulnerabilities exist in a nexus formed by geological and geographical challenges, as they are influenced by a larger socioeconomic and political system. Vulnerability is related to resilience, marginality, susceptibility, adaptability, fragility, and risk (Liverman 1990). Kelly and Adger (2000) argue that people also possess social vulnerability, the capacity of individuals and social groupings to cope, recover from and adapt to external stress on their livelihoods and well-being against socioeconomic and institutional constraints. Fussel (2006), in reviewing the main frameworks of vulnerability, argues that they generally agree on four specific dimensions: system, attribute of concern, hazard, and temporal reference. Within this context, these dimensions correspond to the atoll island system, livelihood security, climate change, and slow onset climate effects. Specific to climate change, the IPCC defines vulnerability as "the propensity or predisposition to be adversely affected. Vulnerability encompasses a variety of concepts including sensitivity or susceptibility to harm or lack of ability to cope and adapt" (IPCC 2014). This explanation considers an outside dimension, which is represented by the 'exposure' of a system to climate variations, as well as an internal dimension, which comprises its 'sensitivity' and its 'adaptive capacity' to these stressors (Fussel and Klein 2006). Additionally, in the scholarship on the human dimensions of climate change, it is recognized that climate change is experienced and responded to in the context of multiple climate and non-climate stressors; assessing island vulnerability to climate change requires understanding how climate conditions are experienced and coped with in the context of other conditions already affecting livelihoods (McCubbin et al. 2015). Climate change multiplies the impact of current stressors on a particular socioeconomic system. Vulnerability, as it relates to livelihoods stress, is what is to be explored further in this chapter. Atoll island livelihoods already experience socioeconomic and environmental stressors apart from climate change, the climate can accelerate and/or exacerbate these stressors. There is no assumption that climate change is a stand-alone factor in climate livelihood stress.

Livelihood vulnerability to climate change can also be usefully understood as an outcome of biophysical and social factors; this refers to the level of exposure communities face from physical impacts of sea level rise, an increase of sea surface and atmospheric temperatures, and the factors that shape the susceptibility of communities to harm and their ability to respond. Climate change increases the vulnerability of rural livelihoods and reduces the ability of households to contend with risks, shocks, and stress (Cutter et al. 2000; Prouse and Scott 2008; Shah et al. 2013). In the case of atoll livelihoods, artisanal fisheries are likely to decline with the increase of coral bleaching and the location of deep water fish may become increasingly unpredictable. Any small-scale agricultural output is also susceptible to damage through increased heat stress, changes in precipitation, saltwater incursion from rising seas, and increased damage from extreme weather events (Barnett and Adger 2003).

How individuals and communities respond to these extraneous environmental shocks is adaptation. Individuals and societies have adapted to climate change over the course of human history and will continue to do so; there have been at similar risks which have acted as drivers for adaptive resource management. Adaptation measures are taken at the individual level, often in response to individual extreme events, while others are taken by governments on behalf of society in response to climate's threat (Adger 2003). This means that there are actions that individuals, households and communities can take alongside those of nations and governments. However, not all actions will be equally effective. Considering socioeconomic and cultural constraints, those individuals with more social and economic capital are more able to afford and to self-adapt than others. This is the same at the nation level; some countries have the economic ability to offer certain protections that others do not. Therefore, adaptation processes involve the interdependence of agents through their relationships with each other, with the institutions in which they reside, and with the resource base on which they depend (Adger 2003). Individuals wield a certain level of agency, even within these institutional and resource constraints, to adjust to the circumstances as necessary. It has been argued that local knowledge is needed for individual level coping and adaptive responses, but in small islands, this base of resilience has been eroded by market modernization (Pelling and Uitto 2001), specifically when (former) subsistence farming and fishing yields are needed to trade in a larger economic marketplace for goods instead of being consumed. However, resource-dependent communities have historically acted collectively to manage weather-dependent, fluctuating, and seasonal resources such as fish, livestock, and water resources on which livelihoods depend. Adger (2003) argues that in this manner, social networks play a primary role in adaptation and recovery. He contends that when governmental intervention to plan for and forewarn communities in disaster planning is absent, social capital replaces the state. However, the level of intervention that a community or individual can do on its' own is minimal.

There are limits at which in situ adaptation may no longer work. When systems change beyond a certain threshold they are often unlikely to retain their original state- the threshold is linked to irreversible change which is a limit on the status quo. However, this is only a relative limit to adaptation in that some adaptation can still be possible when the system moves to another state (Adger et al. 2009). Adger et al. (2009) argue that the discourse around limits to adaptation is frequently constructed around three dimensions: ecological and physical limits, economic limits, and technological limits. After reaching these limits, individual adaptation hinges on whether an impact, anticipated or experienced, is perceived as a risk and whether it should be acted upon. Additionally, many impacts (of climate change) can result in the irreversible loss of assets valued by individuals and sometimes the irreversible damage of asset values. This can include a compromise of capabilities, material and social assets necessary for a means of living (Chambers and Conway 1992). While these are helpful for analytical investigation, there are also individual/household limits- the threshold of conditions at which people can or choose to no longer live with. When this happens, some opt to use a variant of migration.

7.3 The Environment and Migration Decisions

Environmental factors also play a role in migration decisions (Black 2001; EACH-FOR 2009). Research in this area emphasizes that some people who are more exposed to environmental stressors- particularly farmers, herders, pastoralists, and fishermen, may face the greatest obstacles to adaptation (Betts 2010; Black et al. 2011; Warner et al. 2012). Many of these people are exposed to slow onset environmental degradation. Slow onset migration with the possibility of return or permanent displacement is frequently caused by the depletion of resources (land and water), deforestation, desertification, and pollution (Boano et al. 2008). Migration, in this case, is caused by the erosion of resilience, when individuals or communities reach their specific thresholds of their capabilities to mediate damage to their livelihoods or assets. This constrains an agent's ability to make desirable decisions and begins to force them to choose from less preferable options. Thus, the level of access to and control over human, social, cultural, political, economic, and environmental capital characterizes the intensity of exclusion from or inclusion in the process of migration (Kothari 2002). Here, the question of interaction between global climatic change and human migration is not whether environmental drivers are the sole factors causing mobility, but how multiple factors interact to shape migration choices (Warner et al. 2012). Dun and Gemenne (2008) have already argued that migration is a too complex process to simply assume that it can be singularly induced by either the environment or climate in isolation. As mentioned above, livelihood vulnerability is complicated by social, environmental, economic and political systems. The most difficult aspect to unravel from this layered set of circumstances is when and how do people use mobility as an adaptation to climate change? This is what the most recent cross-national research on the topic sought to do. It argues that neither the literature on climate change nor human mobility fully reflects the circumstances under which mobility is an adaptation option. The Rainfalls project[1] directly investigates how changes in rainfall interact with societies to understand the link between changing rainfall patterns, food and livelihood security as well as migration in Bangladesh, Ghana, Guatemala, India, Peru, Tanzania, Thailand, and Vietnam.

The final Rainfalls global policy report finds that 77 % of all households interviewed used migration as a livelihood strategy.[2] However, not all households using migration as a form of adaptation were able to do so effectively. The project divides these households into four categories; (1) those who use migration to improve their resilience (successful migration), (2) those who use migration to survive but are not flourishing, (3) those who use migration as a last resort but is an erosive strategy, and (4) those who cannot migrate, simply struggle to survive, and who are "trapped". While many tend to use migration as a strategy for adaptation based on their households' individual thresholds, not every household can use it in a way that provides successful adaptation and resilience. The Rainfalls report also employs a conceptual framework on which the Rainfalls Agent-Based Migration Model (RABMM) is built, to represent the degree of vulnerability of households to rainfall variability-induced changes in livelihood and food security and the subsequent impact of these on upon the migration of household members (Warner et al. 2012). Ultimately, households need to assess how vulnerable they perceive themselves to be in their current condition and decide if migration is the preferred strategy of adaptation within the given circumstances. This depends on perception of risk, knowledge and experience; choices are shaped by whether local impacts are known and anticipated (Adger et al. 2009). Kothari (2014) reinforces this point by referencing recent empirical work that has also demonstrated that environmental change affects human mobility most directly through livelihoods and the extent to which individuals and communities incorporate risk into their livelihood strategies.

For many, seasonal variability has slowly been experienced and was not previously known or expected. Climate change as a driver for migration is not necessarily obvious; the livelihoods most at risk of slow onset degradation are those in rural areas and individuals may not understand why they are experiencing changes or if they are more widespread than their current region. Thus, migration decisions are also complicated by an insufficient understanding of what is happening on the global level. Additionally, the starting point for much of this literature is the assumption that climate change reduces resources for livelihoods such as food and water (Boano et al. 2008). Most of the research on how climate change will affect islands and atolls points at the degradation of the island's ability to support human habitation as a likely scenario. However, the case study below challenges this supposition.

---

[1] Where the Rain Falls is a collaborative project by CARE France, the United Nations University Institute for Environment and Human Security (UNU-EHS) and the Center for International Earth Science Information Network at the Earth Institute at Columbia University.

[2] This percentage is an aggregate mean of this percentage per location.

The case of Kandholhudhoo's degradation did not follow this general chain reaction of the loss of livelihoods initiating a decision to migrate. The island's inhabitants had a strong livelihood base, which allowed them to cope in situ. However, the island's structural integrity deteriorated under rising seas and most of the inhabitants were not able to adapt via migration, due to socioeconomic and structural barriers. While many studies focus on the potential need for out migration from vulnerable SIDS and atoll nations, what is missing is a focus on specific policy responses to their particular needs in terms of the logistics of migration and a consideration of the barriers that exist, which can prevent migration as a form of adaption and leave many stranded. For islands like Kandholhudhoo, sea level rise posed a temporal threat that was not situated in the distant future. The effects will be sporadic and intermittent currents, tides and storms riding upon a raised sea (Lewis 2007). This means that some locations will incur such damage before others and the process will most likely not have a linear trajectory. This is the reality of Kandholhudhoo and why there is a need for nations such as the Maldives, to consider a development-based internal migration strategy.

7.4 Research Site

The Maldives is a nation of 1192 atoll islands located about 300 miles south of India. Fishing is the mainstay of the traditional Maldivian economy. With only 3000 ha of arable land, most agriculture is subsistence farming mainly native crops such as coconuts, yams and other tree fruits. Additionally, there has been a diversification in the fishing industry away from traditional dry fish productions and towards processing for developed country markets; over half of the total catches of fish are exported (Athukorala 2004). However, tourism is the real driving force of the Maldivian economy, it is the biggest foreign currency earner and main contributor to the GDP. While the unique and isolated nature of the archipelago has helped develop the tourism industry, it has hampered development activities among the dispersed communities beyond fishing and small-scale agriculture (Niyaz and Storey 2011). About half of the human settlements in the country are within 100 m of the shoreline along with almost three quarters of critical infrastructure such as airports, power plants, landfills, and hospitals (Sovacool 2012).

This chapter is based on research conducted in the field in 2013. From October 13 to October 31 of that year, randomly selected semi-structured interviews were conducted on the Island of Dhuvaafaru in Raa Atoll, Maldives. All interviewees were originally from Kandholhudhoo and had been resettled on Dhuvaafaru after the Indian Ocean tsunami of 2004.

This island was chosen as a part of a larger study on perceptions of climate change and migration funded by The Research Council of Norway which included another Maldivian island, K. Guraidhoo, and two additional islands in the Indian archipelago of Lakshadweep. These islands were selected for their similar geographies, religion, economic base, and proximity to India. Interviewees consist of both

Table 7.1 Breakdown of the respondents

Respondent number	Sex	Age	Livelihood
D1	F	38	Housewife
D2	M	43	Fisherman
D3	F	38	Housewife
D4	M	64	Fisherman
D5	F	60	Housewife
D6	F	18	Unemployed
D7	F	28	Unemployed
D8	F	49	Housewife
D9	F	45	Housewife
D10	M	29	Atoll council member
D11	F	64	Housewife
D12	F	24	Housewife
D13	M	Mid 60s	Former island chief/retired
D14	M	74	Retired fisherman
D15	M	56	Security guard for ice plant
D16	F	41	Housewife
D17	F	32	Housewife
D18	F	58	Housewife

Source: Data from Fieldwork in Dhuvaafaru, Simonelli (2014b)

men and women over the age of 18 years. Two hundred quantitative household surveys were conducted and 18 additional people were interviewed in depth[3]; 16 were regular inhabitants with 2 being officials- 1 an elected representative of the Atoll Council and the other the former Island Chief. The interviews were conducted with the assistance of a translator. The breakdown of the respondents was as follows (Table 7.1).

Until 2008, Dhuvaafaru was an uninhabited Maldivian island. It was then that the national government chose it as a relocation site for the inhabitants of Kandholhudhoo, approximately 16–18 km west. The island had to be developed from nothing and building there still continues. The primary economic activity of Dhuvaafaru is fishing; few people work in tourism as there are not many resorts in the atoll. During the field visit, there was only one resort in the atoll and it is approximately 35 min South by speed boat (in good weather) and does not offer any ferry service; any workers must reside on the resort in cramped housing for long periods of time. Another resort on Maamigili, was in the process of being built during the study and opened in early 2014; it is a bit closer to Dhuvaafaru than the current resort but would still be a significant trip by boat. Fishermen often go out for a month at a time leaving the atoll waters to find fish to sell back to their island neighbors and to the capital

---

[3] Some of the in depth interviewees had also completed the survey.

Male. Fishing boats are large and modern, with impressive well–hold capacities. While fishing tends to be considered a 'traditional' livelihood, those continuing to practice it are using every modern tool available. Fishing is also a gendered activity; men become fishermen while women stay home and tend to take care of the wellbeing of their families.

The interviews were informal, usually taking place on the street and a few in people's homes and gardens. Each respondent was asked to discuss the environmental changes that they had observed in any, in their lifetime and to describe them. It became apparent that there was a need to distinguish between their experiences before the tsunami and after as it had a great impact and most individuals began with it unless directed otherwise. This issue was alleviated by asking about their lived experience as it was on Kandholhudhoo and then on Dhuvaafaru separately.

7.5 Study Results

7.5.1 Good Fishing

When discussing their experiences on Kandholhudhoo, several respondents compared their livelihoods between there and Dhuvaafaru unprompted.[4] The respondents offered that the fishing was better in the waters around their old island. Specifically, respondent D9 discussed frequency; when she lived on Kandholhudhoo, the catch was good weekly and now it is good, maybe monthly. D10 and D12 spoke of proximity, now they have to travel much farther away to access the kinds of fish stocks they had been used to catching in the past.

The above description of the status of the fishing is no way an attempt to make a direct comparison of the two islands. The tsunami likely had an effect on the fishing across many islands in the Indian Ocean and the importance of this temporal element has not been overlooked. Additionally, no information was collected as to what the fishing was like in Dhuvaafaru's waters before the tsunami. The comparison is only important to the extent that it establishes that the former inhabitants of Kandholhudhoo perceived that they had a strong economic base during the time that they lived there. "Good fishing" is an indirect indicator of economic strength. As the respondents explained, there was always enough fish to both eat and sell either in local markets or in Male. They also received consistent incomes during this time because their primary livelihood activity was stable and consistent.

---

[4] The semi structured interviews did not specifically ask about fishing and this information was offered independently.

7.5.2 Rising Seas

Of the respondents who were asked,[5] all openly discussed seeing environmental changes in their lifetimes- highlighting the continually rising seas. Many recounted detailed stories of their constant struggle to cope with the sea. Respondent D1 remembers the tides rising higher and higher over time and that they eventually came to scare her. Others, 10 out of the 18 respondents, had experienced water in their homes and at least some damage to their property.[6] This variation was due to how close their home was located to the shoreline. Respondent D9 had water in her house on multiple occasions and did lose some of her belongings, but as she put it, "nothing significant." For respondent D15, a middle aged man, he remembers the tidal flooding becoming especially bad during the monsoons. The water also came into his home and it destroyed some of his kitchen appliances (Respondents D17 and D13 also corroborated that the water was at its worst during the monsoons.) An older woman (D5) recounts having lost household items, such as her television and furniture to the rising waves. Respondent D8 also lost furniture including her bed. For one fisherman (D14), the continued inundation of water almost killed him. He recounts that when the waves came, he used to put sandbags around his house. However, during one incident the water came in and, despite his efforts, rose to be as high as his mid-thigh. He tried to bail water out of his house but eventually lost consciousness in the process and had to be taken to a hospital on a neighboring island.

Beyond personal property damage and the inundation of homes, the constant influx of water affected other facets of the island as well. Respondent D4 recounts that not only did the waves become larger and more destructive over time, but eventually damaged all of the island's trees. Daily activities were also disrupted. One woman explained that when the waves came in all cooking had to stop and the children could not go to school. Additionally, the tides do not simply come in and back out, the flooding would linger for hours.

7.5.3 Coping and Options for Migration

The "good fishing" did allow the interviewees to continue to live on the island because families could afford to replace their belongings over time. This was the continued struggle of those on Kandholhudhoo; loss, community help, and eventual replacement. Respondent, D8, explained that after she had lost most of her household furniture due to the high tides, her family had been in the process of trying to

---

[5] D10, a member of the Atoll Council, did not discuss environmental changes in his interview, but was not asked about them specifically. He was primarily used as reference as to the islands' redevelopment and infrastructure and did not offer any information on this topic independently.

[6] Prior to the tsunami.

replace those belongings at the time of the tsunami. They had not yet been able to when the tsunami hit 6 months later. Her family relied on community help and was not able, even in an economy of "good fishing", to replace such items in half a years' time. Those who had damage to their belongings and sought to repurchase them had to save their money over long periods of time. Referring back to Adger (2003), individuals sometimes take it upon themselves to adapt to extreme circumstances; in the case of Kandholhudhoo, social networks took the place of government assistance. The "good fishing" provided a source of income which served primarily to replace small items lost, help those with bigger losses, and eventually provide a savings base for the replacement of larger items. However, with the damages reoccurring over time, the residents of Kandholhudhoo were beginning to reach the limits of what they could do on their own.

On Kandholhudhoo, the residents experienced years of ever increasing tides and yet most were forced to cope in situ rather than migrate and ultimately can be described as a "trapped population", not unlike those described in the Rainfalls project in category 4. Locals could not migrate due to a lack of personal assets and government help. The inhabitants of Kandholhudhoo faced both socioeconomic and structural barriers to actualizing mobility if it were the preference. Of those interviewed and asked about migrating out of their previous island,[7] only 3 out of 16 had thought seriously about moving but would still be unable to do so without government facilitation.

Respondent D17 considered moving but since there was no government assistance, she said that could not. However, if there had been help she would "have moved to Male." An older man, D15, responded the same. Knowing that there was no government assistance prevented him from considering a move. Most people did not have the capital to do so themselves and without aid from the government, they simply could not migrate. A small reclamation project was enacted by the government in 2002 which restored some of the coast, but as the former Island Chief, D13, explained the East and West sides of the island still flooded badly even after this project and another two to three serious events occurred before the tsunami. He also explained that some individuals[8] did relocate themselves (due to the tidal flooding, before the tsunami) but the government never mandated a large scale relocation. He said that, "Many more talked of moving but could not afford it or did not want to leave friends or family." Government assistance is imperative because there are no other options other than hiring a boat. While many families do own a fishing boat, most are not large enough to move items such as a bed, refrigerator, and the like. Moreover, there is the cost of petrol, the possibility of multiple trips, and taking time away from fishing. And where would one go? If respondent D17 had wanted to move to the capital of Male, it is approximately 165 km (103 mi) away and overcrowded. If those who wanted to move did not have family on Male with whom to stay, the task would be more difficult. The consideration of purchasing land or

---

[7] Neither the Atoll Council member nor the former Island Chief (D10 and D13) were asked personally if they had considered moving from their previous island.

[8] Roughly estimated as a "handful" of families by the former Chief.

building a new home is also an option, but something that is expensive to do if one cannot sell one's home to assist with the financing.

The inhabitants of Kandholhudhoo assisted one another and were able to repurchase their lost items because their economy had not yet been damaged by the other effects of climate change. Each geographical location is unique even though the literature tends to homogenize how climate change will affect small islands. While there is the potential for coral bleaching, and increasing sea temperatures to damage local fisheries, this had not happened yet in Kandholhudhoo; at least to enough of an extent where it interfered with the island's economic backbone. What the IPCC, and other physical scientists cannot say is how larger scale projections will express themselves on the micro level; not every locale will experience every possible climate effect in the same time and in any particular order. Kandholhudhoo was not rendered uninhabitable because its economic base (livelihood) was lost and thus the islanders had no choice but to migrate, their economic base was stable and their island's habitability depended on how often they could recoup their individual losses sustained as their shorelines deteriorated slowly. As far as a threshold is concerned, the continued tidal inundation was enough for those with adequate resources to migrate. But for those without, their agency is curtailed by government non-intervention. Socioeconomically, most could not leave on their own and structurally, those in power provided no alternative to but to remain. Finally, the inhabitants of Kandholhudhoo faced certain types of exclusion to specific sources of capital which led them to be "trapped". As per Kothari (2002), some individuals faced geographical, economic, and political challenges which leads to exclusion. These islanders are geographically remote and isolated without access to services such as publicly funded ferry transport, they faced having to spend their own savings to repeatedly replace items thus depriving them of the ability to sustain an asset base, and finally, with the government deciding that evacuation was not necessary or planning for relocation, they were denied participation and suffered the consequences of a lack of government support.

7.6 Development-Based Migration Policy

Development had always been difficult for the Maldives. Consisting of 7 provinces, 20 atolls, and 194 inhabited islands (Sovacool 2012), political administration is challenging even before one considers development concerns. For every island to be self-sufficient, each would need a harbor, schools, a police station, hospital, waste management, and the staff to run it. Some islands also need water desalination. To provide all of these services to each of the 194 islands is an expensive and monumental task. As Kothari (2014) explains, from the 1980s, it became apparent that a more systematic and widespread proposal for population consolidation was developed, founded on concerns about the diseconomies of scale and the inefficiency of distribution of social services and basic infrastructure on islands with small populations. Over a decade ago, the Safer Island Development Program (SIDP) was developed to identify islands with more favorable geophysical and economic characteristics to make them more resilient to climate change. These were supposed to be safe havens for people forced to migrate before or after natural disasters induced by climate change (Sovacool 2012). This plan was met with considerable resistance (Kothari 2014), and it was in no way helpful for the people of Kandholhudhoo. Even though the program was heavily funded (Sovacool 2012), it was either not given as an option for those on Kandholhudhoo or had not advanced far enough to be an option. It is unclear how far the plan actually developed, but the United Nations Development Program (UNDP) concurred when it reviewed climate change risks throughout the country, noting that no truly "safe" islands exist in the Maldives. This is due to the fact the many homes and infrastructure were designed poorly, sand ridges were leveled during land reclamation and sand mining, coastal vegetation has been destroyed for other land uses, and the islands cannot withstand the rain fall of more severe storms (Sovacool 2012). These are the same problems that befell Kandholhudhoo argues Vince (2009); residents previously dredged the surrounding coral reef for building materials, plundered the mangrove swamps for timber, and felled inland trees that could have substantially mitigated the damage of the tsunami. Because of its lack of defenses, every single house was washed away by a 2.5 m wave.

The SIDP was eventually repackaged by President Nasheed's government as the "Resilient Island" approach through which populations would similarly be encouraged to migrate though incentives of improved social services and transportation networks. Additionally, the government has also embarked on a plan to consolidate the population from almost 200 islands onto 10–15 (Kothari 2014). This plan was mentioned by the Dhuvaafaru Atoll Council representative in the interview. He explained consolidation as a good idea, but it has not been assigned a budget. Additionally, the current government would not allow it.[9] Without a budget, the Maldives had not been able to use development to promote migration to safer islands. One project that has proven popular is the reclamation and redevelopment of the airport island Hulhulmale. Hulhulmale is artificial, or what some call a "designer island" (Sovacool 2012). It is being built at a safer height above sea level with modern services and schools, and close to jobs. In 2013, the government held a raffle to allow some families to move to this island. Many took part, but most did not win the chance to move there (Simonelli 2014b). Sovacool (2012) explains that it currently houses 20,000 but is to being built to house 100,000. Of the participants in the survey who were asked about where they would move if they had to so do again because of further climate impacts, six preferred Hulhulmale and five Male. The reasons they gave were in line with the hopes of the general government conception- that they wanted to be closer to better schools, services, opportunities, and said they thought it to be safe.

Meanwhile, the government has estimated that in 2008, 44 uninhabited islands were under transformation into new tourist resorts and more have been opened

---

[9] The interview was conducted in between a series of several contested elections held after President Nasheed was ousted in a coup. The administration at the time was headed by Mohammad Waheed.

for bidding. Additionally, some inhabited islands were opened up to guest houses (during the Nasheed government) and ten new regional airports were proposed to complement resort development (Niyaz and Storey 2011). As of the field visit in late 2013, a second but much higher-end resort was being built in the Raa atoll but it was unclear when/if a regional airport would be built, but people in the one local resort were hopeful that it would be done. Tourism development, is seems, has far surpassed any large scale projects for Maldivians as a whole, sans Hulhulmale. There is an obvious divide between the two sides of the development coin, private and public. Private development has expanded greatly and is needed in order to fund what can be done in terms of public development. The bed-tax paid by tourists and other taxes related to the hospitality industry have increased over time and will be necessary, however they may not be enough. The Maldives has continued to seek development funds to fill budget gaps and there has been some concern about corruption and where the tourism money actually goes (Simonelli 2014a).

However, policy proposals to draw individuals to new reclaimed islands by economic pull factors can and have shown potential to work, but will need to be more comprehensive to be effective. There is a need to consider current issues as well as future concerns; specifically overpopulation, structural integrity, a limited economic resource base, and temporality. Most inhabited islands are at their limit, with Male- a floating city. Even Dhuvaafaru, after it was constructed for those displaced after the tsunami, is completely full, housing around 5000 people with some houses holding more than one family. Any reclaimed island needs to be large enough to support a growing population, which means leaving enough space for growth instead of filling each island that is built to capacity. Creating artificial islands is feasible and is becoming more frequent. Outside of the Maldives, Malaysia filled in a shallow area to make a tourist resort in the South China Sea, Hong Kong's airport is built over two islands connected by reclamation, and Dubai constructed palm-shaped islands for a luxury residential and leisure area (Kelman 2015). These are examples of private reclamation but demonstrate that large scale constructions are not uncommon- and can be structurally sound; an airport cannot be anything else. However, to continue to build in this fashion under worsening climate effects will necessitate not just building higher, but smarter. This means mangrove protection, importing vegetation, adequate waste management, water catchment (and a possible desalination plant), and sea wall. This will all be very expensive, but Maldivians will be in a continual state of internal migration if their new homes are not built in a manner that is more protective than their previous ones. Dhuvaafaru is higher than the previous island, but its solid waste takes up about 2–3 ha of land in a large pile that is often on fire. If a storm surge hits it, untreated waste will be a large part of the flood waters. This has been problematic in the past and will continue to be in the future if not part of a full scale redevelopment plan.

The limited economic base will be difficult to contend with also. As long as the fishing is good, there will be families that will prefer to continue practicing traditional livelihoods. However, more and more of the young are going to school and will be in search of jobs fitting of an education. If these newly created islands are mini-Males, with a base of local commerce, they will attract young adults who not only want services but city life and their own individual homes. Although the Maldives are an atoll nation of islands, there is an urban/rural divide not unlike any land-locked country. If there are jobs and opportunities on newly developed islands, those who are under duress in far-away atolls will see opportunity and may choose to internally migrate for economic reasons (prior to long term climate stresses) and may inadvertently become safer in the process. Finally, all this will still be temporary. New construction that considers temporality and adaptation will allow more Maldivians to stay in their homes longer. This means finding ways to finance projects that will last the longest under the most uncertainty, and most likely, exploiting the tourism industry for as long as possible. If the resorts are sustainable into the foreseeable future, the government can and will have a solid resource base with which to assist its people.

7.7 Conclusion

In places where the specific geography curtails one's ability to choose migration as adaptation, policy can provide the assistance necessary to allow options that many households could not otherwise choose on their own with their current resource base. However, this necessitates government involvement, sufficient pre planning, and a funding mechanism of some kind. In the Maldives, these policy discussions are not new, but have been moving slowly until recent years. Most concerns surrounding the need for migration as it applies to small islands in the case of climate change tend to be about livelihood continuation as a prerequisite for habitation. However, the effects of climate change will unfold in many ways, and scholars cannot assume that there will be a linear deterioration of islands or homogeneous specific effects. The case of Kandholhudhoo demonstrates that there will be alternative narratives of climate distress, and ones where individual agency, resourcefulness, and resilience can allow for some adaptation in situ. Some individuals were able to use migration as a risk management strategy; they moved early- before their belongings and lives were at stake. But this was not the case for most and thus, there will always be limits to what people can do individually.

For the Maldives, its stable and growing tourism base can provide a macroeconomic fund which, along with development assistance can support development-based incentives to for those in precarious situations to internally migrate. However, doing so requires that there is more than a lottery to allow people to move. Some islanders will be in worse situations than others, due to both structural/economic deterioration and overcrowding. This will have to be dealt with without prejudice because adaptation in the Maldives is effectively on two fronts: adaptation to climate change and to modern development. The overlapping pressures will not be able to be detangled, but can be worked on. Development at the global, meta level has been the cause of climate change and also the pressure to overdevelop the Maldives in ways in which are unsustainable, one of these being tourism. While this is, in essence, a paradox, it is the system in which these issues exist and must be faced. The Maldives' attempt at development-based migration has been semi successful and still has a long way to go, but it is one way in which individuals and families that prefer to stay in their home country can mitigate their exposure to risk and can do so until such time that this also becomes unsustainable. This will depend on how well new policies are enacted and the level of commitment seen from the international community.

References

(2009). EACH-FOR environmental change and forced migration scenarios. In J. Jäger, J. Frühmann, S. Grünberger, & A. Vag (Eds.), *D.3.4 Synthesis report*. European Commission.

Adger, N. (2003). Social capital, collective action, and adaptation to climate change. *Economic Geography, 79*(4), 397–404.

Adger, N., Dessai, S., Marisa, G., Hulme, M., Lorenzoni, I., Nelson, R. D., Naess, L., Wolf, J., & Wreford, A. (2009). Are there limits to adaptation to climate change? *Climatic Change, 93*, 335–354.

Argawal, A., & Perrin, N. (2008). *Climate adaptation, local institutions, and rural livelihoods* (IFRI Working Paper #W081-6). Ann Arbor: School of Natural Resources and Environment, University of Michigan.

Athukorala, P. (2014). Trade policy making in a small island economy: The WTO review of the Maldives. *The World Economy, 27*(9), 1401–1419.

Barnett, J., & Adger, N. (2003). Climate dangers and atoll countries. *Climatic Change, 61*, 321–337.

Betts, A. (2010). Survival migration: A new protection framework. *Global Governance, 16*, 361–382.

Black, R. (2001). *Environmental refugees: Myth or reality?* Geneva: UNHCR.

Black, R., Bennett, S. M., Thomas, S., & Beddington, J. R. (2011). Climate change: Migration as adaptation. *Nature, 478*, 447–449.

Boano, C., Zetter, R., & Morris, T. (2008). *Environmentally displaced people: Understanding the linkages between environmental change, livelihoods, and forced migration*. Oxford: Refugee Studies Centre.

Butler, J. R. A., Skewes, T., Mitchell, D., Pontio, M., & Hills, T. (2014). Stakeholder perceptions of ecosystem service decline in Milne Bay, Papua New Guinea: Is human population a more critical driver than climate change? *Marine Policy, 46*, 1–13.

Chambers, R., & Conway, G. R. (1992). *Sustainable rural livelihoods: Practical concepts for the 21st century* (IDS Discussion paper 296). Brighton: Institute for Development Studies.

Cutter, S., Mitchell, J. T., & Scott, M. S. (2000). Revealing the vulnerability of people and places: A case study of Georgetown County, South Carolina. *Annals of the Association of American Geographers, 90*(4), 713–737.

Dun, O., & Gemenne, F. (2008). Defining 'environmental migration'. *Forced Migration Review, 31*, 10–11.

Fussel, H. M. (2006). Vulnerability: A generally applicable conceptual framework for climate change research. *Global Environmental Change, 17*(2), 155–167.

Fussel, H. M., & Klein, R. J. T. (2006). Climate change vulnerability assessments: An evolution of conceptual thinking. *Climatic Change, 75*, 301–329.

Hare, W. L., Wolfgang, C., Schaeffer, M., Battaglini, A., & Jaeger, C. C. (2011). Climate hot spots: Key vulnerable regions, climate change and limits to warming. *Regional Environmental Change, 11*, S1–S13.

IPCC. (2014). Annex XX: Glossary [J. Agard, E. L. F. Schipper, J. Birkmann, M. Campos, C. Dubeux, Y. Nojiri, L. Olsson, B. Osman-Elasha, M. Pelling, M. J. Prather, M. G. Rivera-Ferre, O. C. Ruppel, A. Sallenger, K. R. Smith, A. L. St. Clair, K. J. Mach, M. D. Mastrandrea, & T. E. Bilir (eds.)]. In Climate Change 2014: Impacts, adaptation, and vulnerability. Part B: Regional aspects. Contribution of Working Group II to the fifth assessment report of the Intergovernmental Panel on Climate Change (pp. 1757–1776) [V. R. Barros, C. B. Field, D. J. Dokken, M. D. Mastrandrea, K. J. Mach, T. E. Bilir, M. Chatterjee, K. L. Ebi, Y. O. Estrada, R. C. Genova, B. Girma, E. S. Kissel, A. N. Levy, S. MacCracken, P. R. Mastrandrea, & L. L. White (eds.)]. Cambridge: Cambridge University Press.

Jäger, J., Frühmann, J., Günberger, S., & Vag, A. (2009). Environmental Change and Forced (EACH-FOR) Migration Scenarios Project Synthesis Report.

Kelly, P. M., & Adger, N. (2000). Theory and practice in assessing vulnerability to climate change and facilitating adaptation. *Climatic Change, 47*, 325–352.

Kelman, I. (2015). Difficult decisions: Migration from small island developing states under climate change. *Earth's Future, 3*(10), 133–142.

Kench, P. S., McLean, R. F., & Nichol, S. L. (2005). New model of reef-island revolution: Maldives, Indian Ocean. *Geology, 33*(2), 145–148.

Kothari, U. (2002). *Migration and chronic poverty* (Working Paper, Vol. 16). Manchester: Institute of Development Policy and Management, University of Manchester.

Kothari, U. (2014). Political discourses of climate change and migration: Resettlement policies in the Maldives. *The Geographical Journal, 180*(2), 130–140.

Lewis, J. (2007). The vulnerability of small island states to sea level rise: The need for holistic strategies. *Disasters, 14*(3), 214–249.

Liverman, D. F., et al. (1990). Vulnerability to global environmental change. In *Understanding global environmental change: The contributions of risk analysis and management*. Worcester: Clark University.

McCubbin, S., Smit, B., & Pearce, T. (2015). Where does climate fit? Vulnerability to climate change in the context of multiple stressors in Funafuti, Tuvalu. *Global Environmental Change, 30*, 43–55.

Niyaz, A., & Storey, D. (2011). Environmental management in the absence of participation: A case study of the Maldives. *Impact Assessment and Project Appraisal, 29*(1), 69–77.

Pelling, M., & Uitto, J. I. (2001). Small island developing states: Natural disaster vulnerability and global change. *Environmental Hazards, 3*, 49–62.

Prouse, M., & Scott, L. (2008). Assets and adaptation: An emerging debate. *IDS Bulletin, 39*(4), 42–52.

Shah, K. U., Dulal, H. B., Johnson, C., & Baptiste, A. (2013). Understanding livelihood vulnerability to climate change: Applying the livelihood vulnerability index in Trinidad and Tobago. *Geoforum, 47*, 125–137.

Simonelli, A. C. (2014a). *Perceptions and understandings of climate change and migration: Conceptualising and contextualising for Lakshadweep and the Maldives* (Field report 1). Norwegian Research Institute.

Simonelli, A. C. (2014b). *Perceptions and understandings of climate change and migration: Conceptualising and contextualising for Lakshadweep and the Maldives* (Field report 2). Norwegian Research Institute.

Sovacool, B. K. (2012). Perceptions of climate change risks and resilient island planning in the Maldives. *Mitigation and Adaptation Strategies for Global Change, 17*, 731–752.

Turvey, R. (2007). Vulnerability assessment of developing countries: The case of small-island developing states. *Development Policy Review, 2*, 243–264.

Vince, G. (2009). Paradise lost? How the Maldives is fighting the rising tide of climate change. *New Scientist, 2707*, 37–39.

Warner, K., & Afifi, T. (2014). Where the rain falls: Evidence from 8 countries on how vulnerable households use migration to manage the risk of rainfall variability and food insecurity. *Climate and Development, 6*, 1–17.

Warner, K., Ehrhart, C., de Sherbinin, A., Adamo, S., & Chai-Onn, T. (2009). *In search of shelter. Mapping the effects of climate change on human migration and displacement.* Chatelaine: CARE International.

Warner, K., Afifi, T., Henry, K., Rawe, T., Smith, C., & de Sherbinin, A. (2012). *Where the rain falls: Global policy report.* Bonn: CARE France and The United Nations University.

Chapter 8
The Reason Land Matters: Relocation as Adaptation to Climate Change in Fiji Islands

Dalila Gharbaoui and Julia Blocher

8.1 Introduction

From early 1990s until today, the low-lying coastal countries known as the Small Island Developing States (SIDS) have progressively gained official recognition and international attention as a group particularly threatened by the adverse impacts of climate change in the coming decades. Land use changes, burning of fossil fuels and greenhouse gas emissions from other sources are exacerbating climate variability and contributing to an increase in global average temperatures. It is likely this warming is compounding the phenomenon of rising sea levels in coastal and low-lying areas as well as the intensifying incidence and magnitude of natural hazards (IPCC 2014a). The Pacific Islands Region (PIR), mainly composed of SIDS, faces increasing threats to survival in the coming decades due to loss of territory and other adverse impacts of climate change. As global climate change mitigation efforts prove insufficient to protect communities and ensure sustainable living conditions, population movements in response to these changes are a reality facing the Pacific islands states today. Such movements include migration, displacement and planned relocations, as agreed, notably, in the outcomes of the 2010 meeting of the parties

---

D. Gharbaoui (✉)
Hugo Observatory for Environmental Migration, University of Liège, Liège, Belgium

Macmillan Brown Center for Pacific Studies, University of Canterbury,
Christchurch, New Zealand
e-mail: dalila.gharbaoui@doct.ulg.ac.be; dalila.gharbaoui@pg.canterbury.ac.nz

J. Blocher
Hugo Observatory for Environmental Migration, University of Liège,
Liège, Belgium

United Nations University, New York, USA

to the UN Framework Convention on Climate Change (UNFCCC) and of the consultation on planned relocation held in San Remo[1] in 2014 (Ferris 2014).

Retreating from coastal areas in response to changing environmental conditions has long played a prevailing role in Pacific Island communities' culture and practices, with land rights at the centre of the process. The fifth assessment report of the Intergovernmental Panel on Climate Change (IPCC) recognized that past examples in the region demonstrate that environmental change can affect land rights and land use, which in turn have become drivers of migration (IPCC 2014a: 1625). Echoing this, the final report of the 3rd International Conference on Small Island Developing States underlined the importance of considering land use and land planning in the context of disaster risk reduction and sustainable development (United Nations General Assembly 2014: 13).

In academic and policy debates on migration as a response to climate change, however, land rights and land tenure systems have been given little importance. It is crucial to address this dimension more in depth, particularly in the context of Pacific Islands, where 80% of land is under customary tenure in the majority of countries (Farran 2011: 65). The main purpose of this chapter is to explore the role of land tenure in environmentally induced planned relocations in the Pacific region by providing a critical evaluation of the different positions developed around "land rights" and "sustainable relocation" and by observing past and recent cases studies in the region. Our ultimate aim will be to provide key insight to the question: to what extent are land rights paramount in forming sustainable adaptive responses to climate change in the PIR? As a corollary, we identify challenges associated with customary land tenure systems and assess how they may be overcome in future planned relocations.

In this framework, this chapter begins by outlining the impacts of climate change and adaptation needs in the Pacific region. The second section considers the concept of sustainable relocation as an adaptive response to climate change impacts and presents current cross-disciplinary debates on relocation strategies, including partial versus staggered schemes. In a third section, the issue of land rights in the Pacific Island Countries and Territories (PICTs) is observed, with an attempt to define the importance of land rights in the sustainability of responses to climate changes. Within this assessment, we engage with literature on the land tenure systems in the region and add to on-going scholarly debates between defenders of collective rights and proponents of individual rights. The foundations of these two camps have implications on how they characterize the relocation process. Emphasis is placed on the neoliberal assumptions associated with land management. Next, case studies will be used in order to explore the role of customary land tenure in past and recent environmentally motivated community relocation in Fiji, in an attempt to draw key lessons for future relocation planning.

---

[1] A distinction is made between temporary relocations (in coherence with international standards), planned relocations (for which relocation is coupled with permanent resettlement, with informed consent) and displacement (forced movements carried out in violation of international standards and spontaneous movements in the wake of natural or manmade hazards, or conflict). See §Ω.1.2.

8.2 Climate Change and the Need for Adaptation

8.2.1 Impacts of Climate Change in the Pacific Islands

Although they are home to a high degree of cultural diversity, all Pacific Islands countries share a number of common features. These include: high dependency on the Pacific Ocean and its natural cycles; a high degree of exposure to natural hazards; fragile natural, human and economic resources; and low capacity to adapt to climate irregularities (SPREP 2009).

The PICTs can be divided in various zones according to their size, resource endowments and the state of economic development (Hughes 2005: 3). Each sub-region; Large High Islands or Melanesian Countries,[2] Mid-Sized High Islands of Polynesia and Micronesia,[3] and Small Islands,[4] is affected differently by the effects of climate change according to its constituents' geographical, ecological or economic features.

Rapid population growth (particularly in urban areas, which are mainly coastal), scarcity of land, land degradation, deforestation, water contamination due to agricultural and mining activities, loss of biodiversity and deterioration of fisheries due to coastal reef decomposition are among the common concerns. Increase in temperatures, precipitation variability and sea level variations can significant affect clean water availability (IPCC 2014a; Zbigniew and Mata 2007: 175; IRIN 2009). Volcanic eruptions, earthquakes and droughts can also have devastating effects.

It is "virtually certain" that the rate of global average sea level rise is accelerating since the 1950s, with the western Pacific experiencing a rise up to four times the global average (IPCC 2014a). Rising sea levels and flooding cause inundation in low-lying areas as well as coastal erosion and sand loss in coastal areas (UNHCR 2010: 8; Nicholls 2003: 6). The fifth assessment report of the IPCC suggested that sea-level rise could pose an existential threat to some atolls islands and will particularly impact islands where communities and infrastructure are located in coastal zones with limited possibility to relocate inland (IPCC 2014a). Although rising sea levels is mainly attributed to warming global average sea temperatures, some non-climate elements should also be considered. Factors such as volcanic activity, the *El Niño* effect· overpopulation (Connell 2003) and sand mining induced coastal erosion (Balinuas 2002) are all also contributors.

Climate change is affecting the essential ecosystems on which the stability of livelihoods, coastal settlements, infrastructure and economic growth in SIDS depends. Most socio-economic activity in the SIDS takes place in low-lying coastal areas (IPCC 2014a: 1625). Sea temperatures affect coral ecosystems and marine

---

[2] Papua New Guinea, Solomon Islands, New Caledonia, Vanuatu, and Fiji.

[3] Tonga, Samoa, French Polynesia, Palau, Federated States of Micronesia, Guam, American Samoa and the Commonwealth of the Northern Mariana Islands.

[4] Cook Islands, Kiribati, Tuvalu, Federated States of Micronesia, the Marshall Islands, Niue, Nauru, and the Republic of Maldives in the Indian Ocean.

life, with serious direct impacts on food security (Fenner et al. 2008; Adams et al. 1996). Changes in fish migration patterns also impact the livelihood habits and food production capabilities of inhabitants, increasing vulnerability (SREP 2009).

There is also a strong correlation between the increasing frequency and intensity of natural hazards such as cyclones, hurricanes, typhoons, and climate change (IPCC 2012: 115). The frequency of natural hazard events has increased in the region since the 1950s, and notably accelerated by 5–20 % in recent years (Bettencourt et al. 2006). Low-lying islands are among the regions of the world most affected by extreme weather events. Recurrent disasters and displacement, sometimes numerous times per year, leave survivors little time to rebuild their lives or means with which to do so. For some, adaptation in situ may not be an option for long. Realistic adaptation strategies – including dimensions that account for the specific economic, social and political contexts of the region – are urgently needed to face an uncertain future.

8.2.2 No Choice But to Move?

Adaptation to the adverse effects of climate change should be coordinated and better mainstreamed with disaster risk reduction and development efforts and must be cautiously planned to avoid increasing risks of vulnerability (IPCC 2014b). The fifth assessment report of the IPCC revealed interesting debates on adaptation opportunities, constraints and limits. Approaching climate adaptation by focusing on risk is particularly useful (Jones and Preston 2011; Dow et al. 2012b) and understanding decisions to relocate through this angle can bring key elements of reflection as it provides a strong ground to determine potential opportunities, constraints, and limits to adaptation (IPCC 2014b).

Moser and Ekstrom (2010) suggest that individual's decision to relocate can take place as "adaptive transformation" to avoid intolerable risk, "[They] may be related to threats to core social objectives associated with health, welfare, security, or sustainability" (Klinke and Renn 2002; Renn 2008; Dow et al. 2013a, b).

At the regional level, relocation has been identified as a main priority for climate change adaptation alongside food security, crop improvement, water security, and resilient infrastructure (SREP 2010: 5; Padma Narsey 2012).

Migration in PICTs, as everywhere, is based on a complex decision-making process involving multiple factors. Barnett and Webber (2010) observe that culture and place attachment are among the most decisive factors (Barnett and Webber 2010: 62). However, the region is noted for its long tradition of environmentally-motivated migration and short-distance whole community relocation (Campbell et al. 2005: 3), and climate change is likely to create long-term and immediate risks that will oblige authorities to relocate the communities facing them.

PICT Governments have started introducing such strategies (AIATSIS 2008). Gradual, facilitated international and internal population relocations have already been suggested within the region. One can point to a number of examples of planned internal relocations executed recently, including a handful within Fiji (Vunidogoloa, Narikoso village and Denumanu Village, Vuya), where 800 coastal and river bank communities are becoming inundated and 45 communities will need to be relocated within the next 5–10 years (Chandra 2015).

Anticipative measures and preparedness strategies in the Pacific, essentially translate to planning for community relocation[5] (Lieber 1977: 343) and rehabilitation[6] (ADB 1998: 3). This chapter defines planned relocation[7] as the planned movement of a population generally conducted with the informed consent of the targeted community and planned by the government, its partners, or a regional entity. It is accompanied by efforts to appropriately compensate the affected population and foster their ability to live sustainably and enjoy rights (de Sherbinin et al. 2010). Relocation may be considered displacement if no such consent and compensation are given, for example, to enable private developers to exploit the land, or following the lawful evacuation of the community because of an imminent threat.[8]

Rehabilitation is critical for all affected communities, whether movement is temporary or permanent, and whether they find themselves in a new or familiar context. Resilience-building efforts must furthermore address the causes of forced migration in the region, including ensuring *sustainable* subsistence of the population and reducing the risk of subsequent displacement. Adaptation measures must be developed and couched within actions that would be taken even in the absence of climate change, due to their contributions to sustainable development.

Three protective dimensions are important in the concept of sustainability, defined at the 1992 United Nations Conference on Environment and Development (UNCED): environmental protection, economic growth and social development (Bruntland Commission 1987; UNCED 2012). Sustainable relocation could therefore be defined as "a process by which a number of {…} people from one locale come to live together in a different locale" (Lieber 1977: 343) whilst simultaneously ensuring protection of the environment, economic growth and social development

---

[5] Lieber defines *community relocation* as: "a process by which a number of … people from one locale come to live together in a different locale", and which requires "rebuilding housing, assets, including productive land, and public infrastructure in another location."

[6] *Rehabilitation* includes "income restoration and re-establishing livelihoods, living, and social systems."

[7] Importantly, this term and type of human mobility was first mentioned in an official text ultimately adopted) of the UNFCCC in 2010, thereby linking the concept to climate change strategies and funding needs in 2010.

[8] At the consultation on planned relocations held in San Remo in 2014, evacuations (the rapid movement of individuals and households – whether advised or mandatory, planned or spontaneous, and conducted in coherence with international standards – to physically protect people from imminent threats) were generally considered as separate from planned relocations (considered as a forward-looking measure in which a whole community is physically relocated *and* permanently resettled). Relocations may, however, follow evacuations in certain circumstances. Displacement, involuntary by definition, may be induced by natural or man-made hazards and also characterizes forced movements carried out in violation of international standards, for example, development-induced displacement.

for the present and without compromising the ability of future generations to do the same.

The next section addresses the concept of sustainable development in relation to relocation strategies mooted in current debates.

8.3 Fostering "Sustainable Relocation"

8.3.1 Relocation Categories and Approaches

Relocation strategies may take various forms and can be organised differently according to varying priorities.

Local relocation or "on-site" relocation (ADB 1998), takes place within national boundaries. In the context of the Pacific, where relocation could be closely related to sea levels in coastal areas, *local relocation* could imply relocating communities to more elevated land positions close to their existing locale (UNHCR 2008: 15).

National relocation takes place within the same country. Both local and National relocation can occur *within* or *beyond* the affected population's *land tenure*[9] boundaries (Campbell et al. 2005: 39) These are critical to the social, economic and political structure of Pacific communities and reflect on "the relationship, whether legally or customarily defined, among people, as individuals or groups, vis-à-vis land" (FAO 2002).

The affected individuals in national and local relocation can be compared to internally displaced persons (IDPs), who share the same legal status as any other citizen and for whom the State has an obligation to respect, protect and fulfil the realisation of rights. A durable solution is defined as a situation in which a displaced person is no longer in need of any specific assistance and protection related to their displacement and can enjoy their human rights without discrimination (IASC 2010).

Regional relocation occurs beyond the national borders but within the region, and still presents an opportunity for affected communities to relocate within similar land tenure systems. For this reason, it has been categorized by some authors as *internal relocation* (within the same land tenure boundaries).

Finally, *international relocation* relates to relocation taking place beyond the national *and* regional boundaries. Proponents of international relocation occasionally moot the responsibility and burden of compensation due to states responsible for historical emissions (c.f. Conisnee and Simms 2003: 36), with limited success.

---

[9] The Food and Agriculture Organisation (FAO) defines land tenure as "an institution, i.e., rules invented by societies to regulate behaviour… They define how access is granted to rights to use, control, and transfer land, as well as associated responsibilities and restraints." (FAO 2002).

8.3.2 Debates Around Staggered Versus Whole Community Relocation

In the debate around so-called *staggered* versus *whole community* relocation, we will detail the perspective of proponents of staggered relocation first. In support of staggered relocation, the statistical relocation approach developed by Zahir, Sarker and Al-Mahmud infers that the relocation strategy should target a selected and restricted portion of the affected community whilst proposing a local adaptation approach for the rest of the affected people (Zahir et al. 2009: 226–350). The argument for this approach is that in focusing on a minimum number of members of the community, it diminishes human and financial costs engendered by full-scale community relocation.

However this statistical approach can be criticised for not taking into account the cultural and sociological elements of the region vis-à-vis community cohesion. Campbell, for example, states that "community disarticulation is arguably the most complex part of the displacement and reconstruction process" (Campbell 2010). The emphasis on social aspects of relocation was already exhorted years before by Perry and Lindell (1997), emphasising the importance of preserving social structure and cohesion by stating that "special attention should be given to social and personal needs of the relocatees {...} (and that) social networks need to be preserved" (Perry and Lindell 1997).

Lieber presents a more historical critique of staggered relocation, concluding from a comparison of ten Colonial-era relocations in the PICTs that preservation of cultural and community cohesion in any relocation of population in the PICTs is paramount (Lieber 1977; UNFCCC 2005: 14).

The aforementioned debates underline the importance of community cohesion in various positions on relocation strategies particularly in the context of the Pacific region. The preservation of community cohesion in the Pacific is strongly dependent on the preservation of land tenure systems, which are central to the social structure of Pacific societies. Land tenure boundaries, discussed in detail below, could therefore serve as the main borders to be considered when planning for relocation of communities.

8.4 The Specifics of Land in the Pacific Islands

Property restitution and access to land are the predominant areas of rights affected by environmental degradation (IASC 2011). Discharging the government's responsibility to protect these rights requires consideration of factors such as existing settlement patterns (plot size of the current and future settlements); land-use habits (current and future settlement basins); and the right to use land (ADB 1998). These factors vary for local, national, regional and international relocations.

8.4.1 Customary Law in the Pacific Islands

Commonalities in the practise of values and customs can be found across the Pacific region (New Zealand Law Commission 2006). Customary laws are central and govern many aspects of Pacific communities' lives, and are essential in building their identity and "world view", conduct of spiritual life, maintenance of cultural heritage and knowledge systems (Tobin 2008).

Customary Law was used as a means of governance in the pre-Colonial era. Zorn (2003) notes that during this period, the legal systems used existing social, political and economic systems to govern the lives of Islanders. Customary laws were functional, had effective methods for ensuring that the rules would be followed and outlined workable procedures for settling disputes (Zorn 2003: 96).

Today, the custom is still the prevailing jurisdictional base for determining the rights, use and boundaries of customary land (Fitzpatrick 2013). Given the importance of land rights to the success of relocations in the PICTs, an understanding of customary law and collective practices is needed as a precursor to planning sustainable relocations.

8.4.2 Land Ownership and Land Tenure

Land tenure refers to societal rules defining how rights to land are allocated, how control and access is granted, specifying conditions and rights for usage, rules for transfer in addition to limitations of use and responsibilities (FAO 2002: 7).

In the PICTs, over 80% of land is ruled under various forms of customary tenure (Farran 2011: 65). Two concepts are of key importance; *Vanua* and *Dela Ni Yavu* (in many Pacific languages). *Vanua* means both the land and the social system and *Dela Ni Yavu* concerns the physical embodiment of the land.

Land tenure systems can both facilitate or provide barriers to relocation according to the context. On one hand, traditional land tenure includes mechanisms for coping with migrants in the PICTs whereby they are accepted on the condition that they become part of the community (Ward 2000: 80). Arrangements can be made between the landholders, land controllers and the newcomers; these typically imply community involvement and asset-based contributions. Whilst rarely formerly recorded, the customs around these agreements are very strong, binding and have long-standing affects. Such customary agreements have facilitated land accessibility to a considerable number of migrants in the PICTs.

On the other hand, land tenure could also be problematic. Despite State claims to property, occupants retain ownership legitimacy via *residual rights* which are customarily linked. Potential allocations of state land relocating communities may result in violent disputes for many years to come, given that the customary land holders are likely to have social and political relationships to the land (Ward 2000: 82). These links form part of their ethnic identity.

It should not be assumed that governments can make independent decisions around land rights and transfer land to relocatees. Consideration of *customary rights holders* is critical given that *overlapping* rights may exist where multiple rights-holders exist over the same land.

It is worth emphasizing that in Pacific Island communities, land is an important source of a sense of security. Land is described as: "an extension of the self; and conversely the people are an extension of the land" (Ravuvu 1988). Customary practices also have critical implications on managing the use of land, forest, and marine resources as well as on food security (AusAID 2008). Collective local practices, inscribed in customary law, are a "shared way of living of a group of people, which includes their accumulated knowledge and understandings, skills and values" (Thaman 2000: 139). As land is the traditional embodiment of islanders' wellbeing, relocation is thus feared to be a potential catalyst for land conflicts and social conflict (Weir et al. 2010: 10).

8.4.3 Debate Over Land Rights: Collective Versus Individual Rights

The diversity of customary land rights in the Pacific region contributed to the complexity of the relocation process. Some stakeholders, ambitious youth and private development projects face contention to replacing collective land rights with individual land rights.

A "non-western" concept of human dignity, among the manifestations of which is communal ownership of land, refers to "societies and institutions that aim to realise human dignity entirely independent of the idea or practise of [individual] human rights" and where *collective rights* predominate (Donnelly 2003: 58). In comparison, international human rights (IHR) are based on "the idea and practice of human rights – equal and inalienable rights held by all individuals against the state and society" (Donnelly 2003: 2).

Scholars argue that collective forms of ownership do not allow individuals to be entrepreneurs, make profit with their land through investment or savings, better attribution of resources, allow for development that improves the overall standard of well-being.[10] Gosarevski et al. (2004) postulate, for example, that customary forms of land tenure are the principal causes of poverty in Papua New Guinea (Gosarevski et al. 2004: 137). Shifting to an individual form of property would purportedly improve living standards, so that poor and marginalized communities can be better included and supported (Gosarevski et al. 2004: 134). Others suggest customary systems have thus far "not been successful" (Lea 2009:63).

Collective forms of land rights may deny certain individual economic and political rights and reinforce inequalities. Customary land owning groups' maintain a

---

[10] Managing social problems, meeting needs, and opportunities for advancement (Midgley 1995: 13–14).

"culture of violence, {...} notably against women" (Gosarevski et al. 2004: 134). Dividends of the system may be counted in social deprivation, marginalization of groups, reproducing discriminations and gender deprivation.[11] These concerns challenge the very notion of international human rights.

"Logistical" approaches to relocation focus on economic growth and development. On the other hand, proponents of collective rights align more with a "sociological approach," supporting internal, cohesive relocation and collective land rights. Proponents of collective rights defend a different understanding of the sense of duty which should be respected (Follesdal 2005; Haines-Sutherland 2009: 2).

8.4.4 Cultural Relativism Versus Cultural Universalism

Cultural relativism[12] is in tension with *universalism*, described by Galtung (1971) as: "the problem of conflicts between Universal Human Rights norms and indigenous social practises that rest on alternative conceptions of human dignity" (Galtung 1971: 83). "Reality" only relates to a social construction in accordance with the prevailing discourse of society (Winch 1964). Human value systems are considered by many to *not* in fact be culturally universal (Ayton-Shenker 1995). The New Zealand Law Commission supports these perspectives arguing that the prevailing concept of Individual Rights "ignores alternative conceptions of rights in non-Western societies, such as Pacific Nations" (NZLC 2006: 31).

Thaman (2000) argues that individual rights are protected by customary law, but are more largely contextualised via collective rights. He asserts that "Pacific societies have long recognized the collective rights of groups and have traditionally protected individual rights in the context of these groups," (Thaman 2000: 393).

Hunt argues that cultural rights have fallen victim to an uneven international system which privileges civil and political rights (Wilson and Hunt 2000: 25). The core conflict is conceived to be tensions in current power relations in the international community, in which the hegemonic Western approach to defending rights is built around the global economy (Smith 2007: 67). Rights are not only culturally specific but politically and economically oriented (Treanor 2004). Notions of group rights in IHR take their origin from an individual standpoint, in the context of "modern" societies.[13] Proponents of collective land rights argue that an imposition of human rights, including individual economic rights, is a politically motivated ideology entrenching the spread of a Universalist vision of economy which does not reflect on the local and cultural specificities of each society.

---

[11] Articles 2(f) and 5(a) of CEDAW require states to modify or abolish customs and practices that constitute discrimination of women (Haines-Sutherland 2010: 131).

[12] Proponents of cultural relativism assert that concepts are socially constructed and vary cross-culturally. These concepts may include such fundamental notions as what is considered true, morally correct, and what constitutes knowledge or even reality itself.

[13] See, for example, article 22 of the Universal Declaration of Human Rights (UDHR) (1948).

In addition, some question the very foundation of the argument that a system based on individual rights actually helps to defend individual economic and political rights. There is little evidence linking the existence of an individual property rights system and increased agricultural productivity and wealth (Gosarevski et al. 2004: 134). Hauffman (2009) argues that a radical deviation from traditional land administration would lead to devastating land poverty (Hauffman 2009: 15).

An intermediate stance exists that advocates for a balance between the two positions. In this scenario, communities choose between traditional forms of land tenure or an individual system (Gosarevski et al. 2004: 135).

The discussion above illustrates that proponents of collective rights have a culturally sensitive focus based on cultural relativist theory. This is in contrast to proponents of individual rights in accordance with IHR theory, a focus based on Universalist theory, who defend merits propagated through economic growth and social development.

Furthermore the debates illustrate the possibility of pursuing an intermediary position that would improve current practices and build on existing customs. This middle ground option is a strong solution for climate-affected communities in the PIR; for relocatees and their host communities alike. In the context of regionally or nationally designed relocations, a middle ground approach would combine logistical and sociological approaches. This would allow PICTs to align with "modern" discourse to deal with the new challenges posed by climate change while involving customary leaders with sensitivity to cultural norms and values. To do so would ensure socio-economic wellbeing by including the population concerned through consultation and negotiation concerning their land rights and options. These notions are further discussed in the concluding section.

8.5 Overview of Recent and Past Examples of Environmentally Induced Community Relocation in Fiji

The section below will provide an illustration of two historical and one recent example of environmentally-induced community relocations with a particular focus on the role that customary land holds in these processes. Potential case studies are abundant in the Pacific region at large: of 86 events involving population movements or relocation of communities[14] assessed by the Asia-Pacific Network for Global Change research (APN), 37 were caused by natural hazards and subsequent disasters, 13 by environmental degradation due to human actions (mining, nuclear testing, inter alia), 9 by conflicts and 9 by development projects (Campbell et al. 2005).

---

[14] Includes community relocations ranging from over 1800 km to less than 1, international and national, under colonial administration and under national sovereign states from 1920 to 2004.

The examples of relocation below took place in the Fiji islands, selected in part because customary land tenure makes up 84 % of the total land area in Fiji.

8.5.1 Biausevu Village

The first example illustrates the community of Biausevu, a village of approximately 150 inhabitants located in the South of Viti Levu Island which was repeatedly relocated between 1881 and 1983 by colonial authorities (Campbell et al. 2005). Relocations occurred following devastating environmental events such as tropical cyclones and heavy rain fall. Initial movement started from Tilivaira to Teagane in 1875, then the community relocated to Biausevu in 1881 and in 1940 from Biausevu to Busadule. The fourth relocation in 1983 was to the elevated site called Koroinalagi (Campbell et al. 2005).

The Biausevu community is linked to several sub-clans (mataqali) and the seat of the high chief of the clan (yavusa). The clan's territory extends around the Biausevu River, including some areas inland and on the coast. The land and its people also called the "Vanua" for Fijians (Ilaitia 2002: 33) is the foundation of the traditional order (Tomlinson 2015) and relates to a communal form of land ownership based on the custom.

A key lesson to be drawn is the positive effect of relocating within the same land tenure boundaries, which simplified access to land and facilitated the relocation process; this may be referred to as "local relocation" (Campbell et al. 2005: 39). The multiple community relocations over 130 years in the Biausevu area still occurred within the same clan's land tenure boundaries and a few kilometres from the initial site (see Fig. 8.1). The role of the community leadership proved critical in negotiating land acquisition, while the support of community-based leaders proved central to the decision making process. As Biausevu was the seat of the clan chief (yavusa), cooperation and assistance to relocation was provided by the surrounding villages belonging to the same social structure. The cooperation of the other communities assisting in resettlement effort was mainly possible thanks to shared leadership structures, culture and traditions, economic ties and land tenure systems. Relocations commenced as close as possible to the community's original home, subsequently moving further away as closer options became intolerable.

However, a number of threats were not alleviated in relocation planning. All attempts were unsuccessful because resettlement areas were as hazardous as the origin areas.

8.5.2 Soldamu Village

The second example of relocation took place in the village of Solodamu, composed of about 100 people divided in 20 families and located on a narrow strip of coastal land on Kadavu Island, the fourth largest island in Fiji. Following a devastating

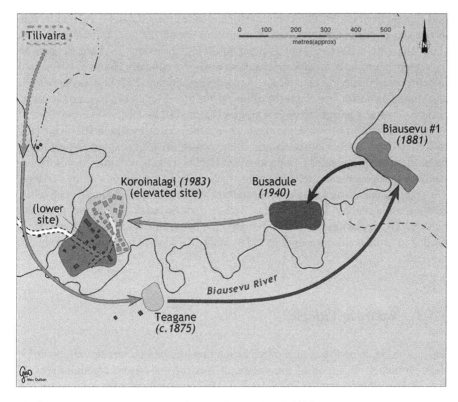

Fig. 8.1 Biausevu relocation project (Source: Campbell et al. 2005)

storm in 1959 and several tropical cyclones in the 1970s that caused the inundation of their home site (Cagilaba 2005), the local chiefs decided to relocate the village onto a hill located 2 km inland. Access to land had to be negotiated, as part of it was owned by the neighbouring village of Natumua. In contrast to the example above, the independent and relatively autonomous communities involved in the Solodamu relocation were not from within the same clan and customary land boundaries. Negotiations to obtain permission to relocate were done according to traditional customs and led by the traditional chiefs (Rokocoko 2006).

On a positive point, the example underlines how traditional methods of negotiation to obtain permission to re-locate conducted between hosting and relocated communities can often allow the relocation to take place smoothly (Rokocoko 2006).

However, the relocation of the Solodamu village has since caused several land access issues with some inhabitants.

Tensions grew over time between the two communities. The host community contested the possibility for the relocatees to extend land limits of the Solodamu new village site. To accommodate their growing population, the Solodamu villagers were compelled to search for other land parcels elsewhere. In addition, the younger generations from the hosting community challenged the relocatees' right to re-side in the site of their ancestors, threatening to re-obtain the land through litigation (Cagilaba 2005). Because land disputes involving customary land are frequent in the region, courts have flexible mechanisms to allow customary claims to be inserted in court cases involving land issues. According to Farran (2011), procedures must accommodate customary claims by allowing "informal" evidence based on hearsay, oral evidence and narratives of genealogies (Farran 2011).

In the Solodamu example, the claim against the relocatees was based on the lack of formal deed. Such conflict underlines the possibility of inter-generational backlashes between proponents of individual land rights (relying on formal deeds to own land) and proponents of collective land rights (relying mainly on oral ownership). The situation underlines need to profoundly consider the long-term protection of the hosting communities' land rights, as well as compensation for the previous land holders. The inclusion of all stakeholders, including inter-generational groups, is needed to avoid conflicts over land and to ensure the long-term security of tenure of relocatees.

8.5.3 *Nakiroso Village*

The last example reveals a case of relocation that took place more recently in 2013; the community relocation of the village of Nakiroso, a coastal community of 27 households on Ono Island in Kadavu in Fiji. The community was forced to relocate as sea level rise had severely affected the area, despite the construction of a rock seawall shoreline in 1960s. The seawall has been gradually eroded, leaving the village under threat of gradual inundation. The first steps of the relocation were carried out by the Fijian government with the support of several ministries, development partners and regional institutions, led by the Secretariat of the Pacific Community (SPC). Special measures ensured active participation from the Narikoso community (SPC 2014). Furthermore, land tenure and its implications were demonstrably considered seriously from the assessment phase, following which a list of recommendations was drawn up in order to facilitate relocation (Pareti 2013). Among the considerations:

Narikoso villagers are furthermore involved in community trainings conducted by the SPC and government experts around integrated farming systems, plant propagation and sustainable land (SPC 2014). The case provides an interesting example of building villagers' capacities to adapt to climate variation and natural disasters, in recognizing that land management and food security are strongly interrelated.

A number of lessons drawn from the examples above could be applied to future relocation processes in the Pacific. A number of positive outcomes merit repetition while certain risks to sustainability of these strategies can be avoided to the extent possible.

The importance of cooperation between the relocating community and the customary land owners of the resettlement site stands out among the points for emulation in the future. Cooperation, collective rehabilitation efforts and longer term community cohesion are often an extension of existing clan ties or trade relationships. On the whole, the sharing of similar institutions of leadership and social structures between origin and destination communities served to facilitate the relocation process. However, conflicts in similar cases are not uncommon. The existence of multiple legal and customary systems today creates vagaries that complicate the decision-making processes around customary land claims. Caution must be exercised when undertaking land negotiations, first to ensure a peaceful process and second to give respect the entrenched practices. Doing so will add to the security of the relocated community's position. Ultimately cooperation as well as integration of communities are possible and can be facilitated through traditional ties. Communities' needs must be more implicated in the post-colonial current context that venerates customary land, given the comprehensive and multi-faceted consideration of all parties, in concertation with all implicated communities and with consideration to environmental and inter-generational effects such as young generations questioning collective land rights of their ancestors. This reality of customary relationships is important to take into account in promoting rights and sustainable development in the region (Campbell et al. 2005).

On the other hand, the case studies demonstrate that care needs to be taken to avoid placing the community somewhere at a similar, or even greater, level of exposure to natural hazards. In today's context, the PICTs can incorporate climate prediction instruments, land use and environmental data as much as possible into such assessments. Historical cases, such as in Biausevu, showed that it might take years to be able to achieve the sustainable resettlement of communities. Relocation as a strategy to adapt to climate change must therefore begin early and in concertation with all implicated communities. In the cases of Narikoso for example, the relocation was initiated by the affected communities that have approached the Fijian government asking for relocation support. It was also the case for the village of Vunidogoloa located in the province of Cakaudrove that was relocated in February 2014 following several decades of coastal erosion that gradually made the site inhabitable. After a lengthy consultation and negotiation process, a site 1.5 km inland and uphill was confirmed in 2012. Here again, the relocation was initiated in 2006 by the head of the Vunidogoloa community that requested support to the Fijian government. There is an urgent need for Pacific governments to anticipate and identify sites most "at risk" in their country in order to start planning relocation as early as possible.

Importantly, the cases demonstrated the extent to which land tenure and land management are central to relocation processes, in both colonial and post-colonial contexts. Customary land tenure and respect for traditional negotiation processes represent both the key to planning and executing a successful relocation as well as the primary risk to the sustainability of the project.

The lessons drawn above should be used to document future relocation policies. Fiji is taking major steps in setting up such policies, which include finalising a relocation guideline in efforts to ensure cooperative and comprehensive treatment of the social, economic and environment issues of the communities concerned (Edwards 2013; Pareti 2013).

The Fijian government is also working together with the Melanesian Spearhead Group to develop appropriate strategies to address environmentally induced migra-

Fig. 8.2 Vunidogoloa Relocation Project (2014) (Source: Hannah Entwisle (Nansen Initiative))

tion at regional level (Fiji National report 2013). In the next 5–10 years, 34–45 villages are expected to be relocated as result of coastal erosion and sea level rise (RTCC 2014; Chandra 2015). The confirmation of which villages are to be relocated will be possible after necessary assessments are finalized, however, it appears that one of the next projects will be to relocate Waciwaci District School in Lakeba, an island in Fiji's southern archipelago of Lau (Chandra 2015). In those future relocation projects land and conflicts associated to cultural patterns should be considered as extremely sensitive. Colonel Apakuki Kurusiga, Deputy Head of the Ministry of iTaukei Affairs, confirms: (Fig. 8.2).

Caption: Government-supported due to the encroaching sea, the case of Vunidogoloa is considered to be the first community relocation in Fiji related to climate change-induced sea level rise. The villagers relocated to Vanua Levu, Fiji's second largest island, not more than 2 km from their old village site but on higher ground. The new site has approximately 30 houses, each equipped with its own solar panel, has ponds for fish and prawn production, cattle, coconut farms, and a new community building under construction.

> *For generations, a community has come to identify with the piece of land they have called home, and in these they have stored their history, their genealogy and their very being {…}. So it is by no means an easy exercise for these coastal communities to leave their "yavu" and relocate* (Pareti 2013).

8.6 Conclusion

The forecasted impacts of climate change bring unprecedented policy challenges to small island states. For these communities, opportunities to adapt in situ to environmental and meteorological threats are in most cases increasingly scarce. As relocation will become increasingly attractive as a strategy to deal with the risks of climate change, numerous considerations must be made to ensure population movements are effective in the long-term. This chapter sought to lay the groundwork to deliver the potential of communities to ensure planned relocations are sustainable. We chose to focus on the Pacific due to specificities and disputes that arise within the customary or the Western-introduced systems of tenure as well as in the interaction of both systems. However, many of the points elaborated above can be applied to other relocations, particularly those in which communication between relocating and host communities are paramount. Accommodating migration for individuals and communities compelled to move implies new conditions of land use, creating new and unique sensitivities. Relocation can be a successful risk management strategy for communities and the individuals they are composed of; however, steps can be taken to mediate resulting and unintended risks.

The UN definition of sustainable development – focusing on social, economic, and environmental aspects – was taken as a point of departure to assess the criteria for success of planning relocations for climate-affected communities. While this definition proved insufficient for many socio-cultural concepts familiar to Pacific communities, it framed our exploration of various viewpoints on the collective notion of land tenure and rights versus its individual based counterpoint. In critically engaging in this scholarly debate, we dissected current discussions on relocation strategies: arguments around social development, community well-being based on land cohesion, the preservation of collective land rights and "cultural relativism" on one hand, and neoliberal arguments on economic growth, the promotion of individual land rights and logistical aspects of the relocation process on the other hand. The collective rights model described above is not coherent with the capitalist model of productivity and economic growth. Each model follows a different conception of accumulation of "wealth," land ownership and values. In questioning the neoliberal conception of land rights, we expose a counterpoint based on non-commercial uses of resources.

Our discussion exposed a crucial avenue to be assessed in regards to ensuring the long-term success of relocations, of socio-cultural aspects in particular. We show that consensual notions of land management in market approaches to land value is artificial and inappropriate in the context of the Pacific region where the concept of communal ownership prevails over the notion of individual responsibility.

Ultimately, we propose an intermediate position, reflecting on the possibility for the two approaches identified above to be complementary. This middle ground entails elements of cultural relativism along with defenders of aligning the Pacific to neoliberal doc-trines of land management as a means to deal in an efficient way with new challenges posed by climate change. This would translate in a combined 'sociological and logistical' approach that includes negotiation at early stages of the relocation process including governments, local leaders and both relocatees and hosting communities. Inserting local adaptation strategies and traditional knowledge on environment in regional and national decision-making is crucial, however, it is important to also insert measures supporting awareness raising on climate change amongst customary high chiefs in parallel as "scientific ideas and theories about environmental protection could be perceived as luxury that contradicts an established set of life-sustaining practices on the island" (Worliczek & Allenbach 2011).

With case studies of environmentally-induced relocations in Fiji, we demonstrated that long-held customs as well as shared institutions and structures can be key to facilitating relocations and maintaining positive ties that may adapt to contemporary needs. We showed that relocating within the boundaries of a common customary land tenure system simplifies the relocation process because of considerations of land tenure system, leadership structures and willingness of communities to cooperate. At the same time, communities' leadership in the negotiation over land at the assessment phase and throughout the relocation process is key to sustainable relocation. Our case studies confirm that consultation processes around relocation should include all interests at stake (community leaders, relocated and hosting communities) and find a way to combine both neo liberalist conceptions and collective consideration of land management in the region. This combined approach would therefore also inform on potential inter-generational tensions, which in practice can mean conflict between proponents of individual land rights contrasted with proponents of collective land rights; one case study revealed the importance of considering inter-generational effects of customary tenure. Appropriate and culturally sensitive forms of land management should be inserted in international and regional governance of climate change and migration, in order to avoid a mismatch between real-life processes and normative perspectives in regions where communal notion of land coexists with the expanding globalization and neo-liberalization.

Finally, we note that following migration, insecurity of land tenure could potentially be very high. Clear government policies on climate change adaptation and relocation planning which combine both modern and traditional approaches, along with ensured access to reliable information for affected communities, are paramount.

Applying an intermediate position to relocation beyond neoliberal discourses on land management is crucial. Without deep exploration of ancestral and recent community relocation in the region by taking into account lessons on the implications of customary land tenure in this process, future planning on relocation is likely to be unsustainable. This is because it will fail to include the essential link between Islanders and their land, an extension of their identity for millennia. All stakeholders including civil society, development programmes, PIR decision-makers and international agencies must work together to find ways to conduct efficient and technically realisable population relocation strategies, while at the same time considering the needs of the community in allowing them to preserve their identity, social and belief systems, and community cohesion. The middle-ground solution proposed above conceptualized as "the middle-ground approach to relocation", is likely to mediate to the greatest extent possible a wide range of potential risks to the long-term success of relocations. Furthermore, it provides a basic structure that may be replicated for a number of cases that involve land or resources disputes between communities affected be environmental threats.

The major challenge will be to pursue those considerations despite an unfortunate and reactionary preoccupation with crises. This was recently illustrated by aftermath of Cyclone Pam in Vanuatu on 13th March 2015, "A disaster of this magnitude has not been experienced by Vanuatu in recent history – particularly in terms of the reach of the potential damage and the ferocity of the storm," resulting in the displacement of 3300 persons (OCHA 2014 [n.p]).

We conclude by advocating for further research on the interface between state-based ("formal") and community-based ("indigenous") land security mechanisms as well as on the implications of land reform and land-based conflicts in planning for migration as adaptation strategies in the Pacific region. The evolution of the land reform and the future survival of customary land rights as prevailing form of tenure in the Pacific will have significant incidence on future policies related to land management and environmental migration in the region.

References

Adams, T., Dalzell, P., & Farman R. (1996). Status of pacific island coral reef fisheries. 8th international coral reef symposium. Panama: SPC Coastal Fisheries Programme, Service de la Mer, Province Sud, Noumea.

Alchian, A. (2008). Property rights. In *The concise encyclopedia of economics*. 2008. Library of economics and liberty. http://www.econlib.org/library/Enc/PropertyRights.html. Accessed 25 Feb 2016.

Anderson, T. (2006). On the economic value of customary land in Papua New Guinea. *Pacific Economic Bulletin, 21*(1), 2006.

Asian Development Bank (ADB). (1998). *Handbook on resettlement: A guide to good practice* [Online]. http://adb.org/sites/default/files/pub/1998/Handbook_on_Resettlement.pdf. Accessed 25 Feb 2016.

Asian Development Bank (ADB). (2000). Land tenure in the pacific islands: Changing patterns and implications for land acquisition. In R. Manning (Ed.), *Resettlement policy and practice in Southeast Asia and the Pacific* (pp. 75–87). Manila: Asian Development Bank.

Asian Development Bank (ADB). (2012). *Adressing climate change and migration in Asia and the Pacific*. Madaluyong City: Asia Development Bank.

Australian Agency for International Development (AusAID). (2008). *Making land work: Case studies on customary land and development in the Pacific* (Vol. 2). Canberra.

Australian Institute of Aboriginal and Torres Strait Islander Studies (AIATSIS). (2008). *Impacts and opportunities of climate change: Indigenous participation in environmental markets* (Land, rights, laws – Issues of native , Vol. 3, Issue Paper No. 13, Native Title Research Unit, p. 3). http://www5.austlii.edu.au/au/journals/LRightsLaws/2008/2.pdf. Accessed 25 Feb 2016.

Ayton-Shenker, D. (1995). *The challenge of human rights and cultural diversity*. United Nations Background Note, United Nations Department of Public Information.

Balinuas, S., & Soon, W. (2002). Is Tuvalu really sinking? *Pacific Magazine, 28*(2), 44–45.

Barnett, J., & Webber, M. (2010). *Accommodating migration to promote adaptation to climate change* (Policy Research Working Paper, Vol. 5270, p. 62). Washington, DC: The World Bank.

Bell, D. (1996). The East Asian challenge to human rights: Reflections on east west dialogue. *Human Rights Quarterly, 18*(3), 641–667.
Bettencourt, S., Croad R., Freeman P., Hay, J., Jones, R., King, P., Lal, P., Mearns, A., Miller, G., Psawaryi-Riddhough, I., Simpson, A., Teuatabo, N., Trotz, U., & van Aalst, M. (2006). *Not If but when: Adapting to natural hazards in the pacific islands region: A policy note*. The World Bank, East Asia and Pacific Region, Pacific Islands Country Management Unit, Washington DC, p. 43. http://siteresources.worldbank.org/INTPACIFICISLANDS/Resources/Natural-Hazards-report.pdf. Accessed 25 Feb 2016.
Biermann, F., & Boas, I. (2007, November). *Preparing for a warmer world: Towards a global governance system to protect environmental refugees* (Global Governance Working Paper No 33).
Brundtland Commission. (1987). Report of the World Commission on Environment and Development. United Nations 1987.
Burson, B. (Ed.). (2010). *Climate change and migration-south pacific perspectives*. Wellington: Institute of Policy Studies. http://ips.ac.nz/publications/files/6666ee71bcb.pdf. Accessed 25 Feb 2016.
Cagilaba, V. (2005). *Fight or flight? Resilience and vulnerability in rural Fiji*. Department of Geography. Hamilton: University of Waikato. p. 123.
Campbell, J. R. (2006). *Traditional disaster reduction in Pacific Island communities*, GNS Science Report 2006/38 p. 46.
Campbell, J. R. (2010). An overview of natural hazard planning in the Pacific Island Region. *The Australasian Journal of Disaster and Trauma Studies*, ISSN: 1174-4707, vol. 2010–2011. http://www.massey.ac.nz/~trauma/issues/2002-2/mcdowell.htm. Accessed 25 Feb 2016.
Campbell, J. R., Goldsmith, M., & Koshy, K. (2005). *Community relocation as an option for adaptation to the effects of climate change and climate variability in Pacific Island Countries (PICs)*. Kobe: Asia-Pacific Network for Global Change Research (APN). Final report for APN Project 2005-14-NSY. http://www.pacificdisaster.net/pdnadmin/data/original/APN_2009_PICS_community_relocation.pdf. Accessed 25 Feb 2016.
Chandra, T. (2015). 45 Fiji Villages likely to be relocated due to climate change. *Pacific Islands Report*, Pacific Islands Development program (PIDP), Fiji Times. http://pidp.eastwestcenter.org/pireport/2015/February/02-13-03.htm. Accessed 25 Feb 2016.
Conisbee, M., & Simms, A. (2003). *Environmental refugees: The case for recognition*, New Economics Foundation. http://www.goodplanet.info/goodplanet/index.php/eng/Contenu/Points-de-vues/Eco-refugies-une-lutte-pour-la-reconnaissance/(theme)/289. Accessed 25 Feb 2016.
Connell, J. (2003). Losing Ground Tuvalu, the greenhouse effect and the garbage can. *Asia-Pacific View Point, 44*(2), 89–107.
De Sherbinin, A., Gemenne, F., & Marcia, C. (2010). *Preparing for population displacement and resettlement associated with large climate change adaptation and mitigation projects*. Background Paper for the Bellagio Workshop 2–6 November 2010. http://www.ciesin.columbia.edu/confluence/download/attachments/92799210/Background_Paper_final.pdf. Accessed 25 Feb 2016.
Donnelly, J. (2003). *Universal human rights in theory and practice* (2nd ed.). Ithaca: Cornell University Press.
Dow, K., Berkhout, F., & Preston, B. L. (2013a). Limits to adaptation: A risk approach. *Current Opinion in Environmental Sustainability, 5*(3–4), 384–391.
Dow, K., Berkhout, F., Preston, B. L., Klein, R. J. T., Midgley, G., & Shaw, R. (2013b). Limits to adaptation. *Nature Climate Change, 3*, 305–307.
Edwards, J. (2013, January). *Climate-induced relocation: A first for Fiji*. http://www.methodist.org.uk/media/757401/wcr-pcc-relocation-policy-fiji-jan13.pdf. Accessed 25 Feb 2016.
Engle Merry, S. (2006). *Human rights and gender violence: Translating international law into local justice*. Chicago: University of Chicago Press.
Etzioni, A. (1997). The end of cross-cultural relativism. Alternatives. *Social Transformation and Humane Governance, 22*(2), 177–189.

Euractiv. (2002). *Sustainable development: Introduction.* http://www.euractiv.com/sustainability/sustainable-development-introduction/article-117539. Accessed 25 Feb 2016.

Farran, S. (2011). Navigating between different land tenure and land laws in pacific island states. *Journal of Legal Pluralism* 64. http://commission-on-legal-pluralism.com/volumes/64/farran-art.pdf. Accessed 25 Feb 2016.

Fenner, D., Speicher, M., Gulick, S. [coords.] (2008). *The state of coral reef ecosystems of American Samoa.* http://americansamoa.noaa.gov/pdfs/coral_ecosystem.pdf. Accessed 25 Feb 2016.

Ferris, E. (2014). *Planned relocations, disasters and climate change: Consolidating good practices and preparing for the future.* United Nations High Commissioner for Refugees (UNHCR) -Brookings-Georgetown 2014). Background document to the San Remo Consultation. http://www.unhcr.org/53c4d6f99.pdf. Accessed 25 Feb 2016.

Fiji Government. (2013). On progress in implementation of the Mauritius Strategy for Further Implementation (MSI) of the Barbados Programme of Action (BPOA) 2010–2012, Fiji National Report, June 2013. http://www.sids2014.org/content/documents/218Fiji%20report.pdf. Accessed 25 Feb 2016.

Fingleton, J. (2007, March). Rethinking the need for land reform in Papua New Guinea. *Pacific Economic Bulletin, 22*(1).

Fitzpatrick, D. (2013). Land and human mobility in the pacific: The effects of natural disasters. discussion paper. *The Nansen Initiative*: 16. https://www.nanseninitiative.org/pacific-consultations-intergovernmental/. Accessed 25 Feb 2016.

Fletcher, S. M., Thiessen, J., Gero, A. Rumsey, M., Kuruppu, N., & Willetts J. (2013). Traditional coping strategies and disaster response: Examples from the South Pacific Region. *Journal of Environmental and Public Health* 2013, Article ID 264503, 9 pages http://dx.doi.org/10.1155/2013/264503. Accessed 25 Feb 2016.

Follesdal, A. (2005). Human rights and relativism. In A. Follesdal, & T. Pogge (Eds.), *Real world justice: Grounds, principles, human rights standards and institutions.*

Food and Agriculture Organization of the United Nations (FAO). (2002). *Land tenure and rural development.* FAO Land Tenure Studies 3, FAO Corporate Document Repository, Rome 2002. http://www.fao.org/docrep/005/y4307e/y4307e05.htm. Accessed 25 Feb 2016.

Food and Agriculture Organization of the United Nations (FAO). (2008). *Climate change and food security in Pacific Island Countries. Issues and requirements.* http://www.fao.org/docrep/011/i0530e/i0530e00.HTM. Accessed 25 Feb 2016.

Galtung, J. (1971). A structural theory of imperialism. *Journal of Peace Research, 2*, 81–117. http://jpr.sagepub.com/cgi/pdf_extract/8/2/81. Accessed 25 Feb 2016.

Gerrard, M. B., & Vannier, G. E. (2015). *Threatened island nations: Legal implications of rising seas and a changing climate.* Cambridge: Cambridge University Press.

Gosarevski, S., Hughes, H., & Windybank, S. (2004). Is Papua New Guinea viable? *Pacific Economic Bulletin, 19*(1), 134–148.

Guterres, A. (2009). *Five 'mega-trends'—Including population growth, urbanization, climate change—Make contemporary displacement increasingly complex, third committee told, General Assembly GA/SHC/3964.* New York: United Nations.

Haines-Sutherland, K. (2009). *Balancing human rights and customs in the Pacific Region.* New South Wales: Australian National University. http://www.cla.asn.au/Article/2009/Paper%20Kelly%20HS%200912.pdf. Accessed 25 Feb 2016.

Haines-Sutherland, K. (2010). Balancing human rights and customs in the Pacific Region: A pacific charter of human rights? *The ANU Undergraduate Research Journal, 2,* 2010. https://eview.anu.edu.au/anuuj/vol2_10/pdf/ch08.pdf. Accessed 25 Feb 2016.

Hauffman, K. (2009, February 9). A palmy balm for the financial crisis. *Sydney Morning Herald.*

Hughes, A. V. (2005). *Strengthening regional management a review of the architecture for regional co-operation in the Pacific.* Consultative draft report to the Pacific Islands Forum. http://www.sopac.org/sopac/docs/RIF/06_AV%20Hughes%20Report_CONSULTATIVE_DRAFT%281%29.pdf. Accessed 25 Feb 2016.

Humanitarian Affairs (OCHA) (2014). Fiji: Building resilience in the face of climate change [Web]. http://www.unocha.org/top-stories/all-stories/fijibuilding-resilience-face-climate-change. Accessed 25 Feb 2016.

Ilaitia, S. T. (2002). *Vanua: Towards a Fijian theology of place*. Institute of Pacific Studies, University of the South Pacific, p. 245

Inter-Agency Standing Committee of the United Nations (IASC). (2010). *IASC framework on durable solutions for internally displaced persons*. http://www.unhcr.org/50f94cd49.pdf. Accessed 25 Feb 2016.

Inter-Agency Standing Committee of the United Nations (IASC). (2011*). Operational guidelines on the protection of persons in situations of natural disasters*. The Brookings-Bern project on internal displacement. http://www.ohchr.org/Documents/Issues/IDPersons/OperationalGuidelines_IDP.pdf Accessed 25 Feb 2016.

Intergovernmental Panel on Climate Change (IPCC). (2001). *Working Group II: Impacts, adaptation and vulnerability*. http://www.ipcc.ch/ipccreports/tar/wg2/index.php?idp=671. Accessed 25 Feb 2016.

Intergovernmental Panel on Climate Change (IPCC). (2007). *Climate Change 2007: Synthesis report* (p. 104). Contribution of Working Groups I, II and III to the Fourth Assessment Report of the Intergovernmental Panel on Climate Change (Core Writing Team, Pachauri, R. K. & Reisinger, A., Eds.). Geneva: IPCC.

Intergovernmental Panel on Climate Change (IPCC). (2012). In C. B. Field, V. Barros, T. F. Stocker, D. Qin, D. J. Dokken, K. L. Ebi, M. D. Mastrandrea, K. J. Mach, G.-K. Plattner, S. K. Allen, M. Tignor, & P. M. Midgley (Eds.), *Managing the risks of extreme events and disasters to advance climate change adaptation* (A Special Report of Working Groups I and II of the Intergovernmental Panel on Climate Change, p. 582). Cambridge, UK: Cambridge University Press. https://www.ipcc.ch/pdf/special-reports/srex/SREX_Full_Report.pdf. Accessed 25 Feb 2016.

Intergovernmental Panel on Climate Change (IPCC). (2014a). Chapter 29: Small islands. http://ipcc-wg2.gov/AR5/images/uploads/WGIIAR5-Chap29_FGDall.pdf. Accessed 25 Feb 2016.

Intergovernmental Panel on Climate Change (IPCC). (2014b). Chapter 16: Adaptation: Opportunities, constraints and limits. http://www.ipcc.ch/pdf/assessment-report/ar5/wg2/WGIIAR5-Chap16_FINAL.pdf. Accessed 25 Feb 2016.

IRIN News. (2009). *Climate change's threat to water needs more study*. http://www.irinnews.org/Report.aspx?ReportId=81034#. Accessed 25 Feb 2016.

Jones, R. N., & Preston, B. L. (2011). Adaptation and risk management. *Wiley Interdisciplinary Reviews: Climate Change, 2*, 296–308.

Klinke, A., & Renn, O. (2002). A new approach to risk evaluation and management: Risk-based, precaution-based, and discourse-based strategies. *Risk Analysis, 22*, 1071–1094.

Kundzewicz, Z. W., Mata, L. J., Arnell, N. W., Döll, P., Kabat, P., Jiménez, B., Miller, K. A., Oki, T., Sen, Z., & Shiklomanov, I. A. (2007). Freshwater resources and their management. Climate Change 2007: Impacts, adaptation and vulnerability. In M. L. Parry, O. F. Canziani, J. P. Palutikof, P. J. van der Linden, & C. E. Hanson (Eds.), *Contribution of Working Group II to the fourth assessment report of the Intergovernmental Panel on Climate Change* (pp. 173–210). Cambridge: Cambridge University Press. http://www.ipcc.ch/pdf/assessment-report/ar4/wg2/ar4-wg2-chapter3.pdf. Accessed 25 Feb 2016.

Lea, D. (2009). New initiatives in the development of customary land: Group versus individual interests. *Pacific Economic Bulletin, 24*(1), 2009.

Lieber, M. D. (1977). *Exiles and migrants in oceania* (Association for Social Anthropology in Oceania Monograph Series). Honolulu: The University Press of Hawaii.

Lopez, A. (2007, March 22). The protection of environmentally-displaced persons in International law. *Environmental Law*. http://www.thefreelibrary.com/The+protection+of+environmentally-displaced+persons+in+International+...-a0165017867. Accessed 25 Feb 2016.

Midgley, J. (1995). *Social development: The developmental perspective in social welfare.* London: Sage. A discussion of the social development approach to harmonizing economic and social policies in the development process.

Moser, S. C., & Ekstrom, J. A. (2010). A framework to diagnose barriers to climate change adaptation. *Proceedings of the National Academy of Sciences of the United States of America, 107*(51), 22026–22031.

Nansen Initiative. (2011). *Conclusion: Nansen Initiative Pacific Regional Consultation.* http://www2.nanseninitiative.org/pacific-consultations-intergovernmental/. Accessed 25 Feb 2016.

New Zealand Law Commission (NZLC). (2006). *Converging currents: Custom and human rights in the pacific* (Study Paper 17). http://ir.canterbury.ac.nz/bitstream/handle/10092/3372/12610366_converging_currents.pdf?sequence=1&isAllowed=y. Accessed 25 Feb 2016.

Nicholls, J. (2003). Case study on sea-level rise impacts, Working party on global and structural policies; OECD workshop on the benefits of climate policy: Improving information for policy makers. http://www.oecd.org/dataoecd/7/15/2483213.pdf. Accessed 25 Feb 2016.

Office for the Coordination of Humanitarian Affairs (OCHA) (2014). *Fiji: Building resilience in the face of climate change [Web].* http://www.unocha.org/top-stories/all-stories/fiji-building-resilience-face-climate-change. Accessed 25 Feb 2016.

Padma Narsey, L. (2012). *Climate change adaptation in the Pacific: Making informed choices. Summary for decision makers.* http://www.pacificdisaster.net/m/nav_organisations.jsp?parent=AusAID#show_article.jsp?id=15721. Accessed 25 Feb 2016.

Pareti, S. (2013, May). Climate refugees? More coastal villages may have to relocate due to eroding shorelines and coastal flooding. *Island Business.* http://www.islandsbusiness.com/2013/5/fiji-business/climate-refugees/ .Accessed 25 Feb 2016.

Perry, R. W., & Lindell, M. K. (1997). Principles for managing community relocation as a hazard mitigation measure. *Journal of Contingencies and Crisis Management, 5*(1), 53–56.

Ravuvu, A. D. (1988). Development or dependence: The pattern of change in a Fijian Village, Suva, Institute of Pacific Studies and Fiji Extension Centre, University of the South Pacific.

Renn, O. (2008). *Risk governance: Coping with uncertainty in a complex world* (p. 368). London: Earthscan.

Responding to Climate Change (RTCC). (2014). *Fiji village relocated under climate change programme.* http://www.rtcc.org/2014/01/17/fiji-village-relocated-under-climate-change-programme/#sthash.dbj8Y9XN.dpuf. Accessed 25 Feb 2016.

Rokocoko, V. (2006). Fight and flight: Challenges faced by two Fijian villages adapting to climate change. *Oceanic Waves,* University of South Pacific, Suva, Fiji, 8(4).

Secretariat of the Pacific Community (SPC). (2014). Nakiroso villagers elated with first harvest since planned relocation, 21 August 2014, Suva. http://www.spc.int/en/media-releases/1796-narikoso-villagers-elated-with-first-harvest-since-planned-relocation.html

Secretariat of the Pacific Regional Environment Programme (SPREP). (2009). Pacific adaptation to climate change cook islands, report of In-Country Consultations. https://www.sprep.org/attachments/67.pdf. Accessed 25 Feb 2016.

Smith, K. M. (2007). *Textbook on international human rights* (3rd ed.). Oxford: Oxford University Press.

Solomon Times Online. (2009). Kiribati seeks relocation as climate change sets in. http://www.solomontimes.com/news.aspx?nwID=3584. Accessed 25 Feb 2016.

Stege, K. E., & Huffer, E. (2008). Land and women: The matrilineal factor: The cases of the Republic of the Marshall Islands, Solomon Islands and Vanuatu Pacific Islands Forum. Secretariat | Triennial Pacific Conference of Women 2007: Nouméa.

Thaman, K. H. (2000). Cultural rights: A personal perspective. In M. Wilson & P. Hunt (Eds.), *Culture, rights and cultural rights: Perspectives from the South Pacific.*

Tobin, B. (2008). The Role of Customary Law in access and benefit-sharing and traditional knowledge governance: Perspectives from Andean and Pacific Island Countries, UNU- WIPO. www.wipo.int/export/sites/www/tk/en/resources/pdf/customary_law_abs_tk.pdf. Accessed 25 Feb 2016. In Customary Law and Traditional Knowledge, Background Brief 7, p.1. http://www.wipo.int/edocs/pubdocs/en/wipo_pub_tk_7.pdf. Accessed 25 Feb 2016.

Tomlinson, M. (2015). Gender in a land-based theology. *Oceania, 85*, 79–91.
Treanor, P. (2004). *Why human rights are wrong.* http://web.inter.nl.net/users/Paul.Treanor/human-rights.html. Accessed 25 Feb 2016.
Tsuchida, Y. (2008). Climate change, natural disasters and human displacement, International conference on Environment, Forced Migration and Social Vulnerability (EFMSV), Bonn, 10 October 2008 *In.* UNCHR, http://www.unhcr.org/4901e9bd2.pdf. Accessed 25 Feb 2016.
United Nations Conference on Environment and Development (UNCED). (2012). Declaration of Rio on Environment and Development. In United Nations Environment Programme, *Rio Declaration on Environment and Development.*
United Nations Environment Programme. (2007). IPCC fourth assessment report (AR4), Valencia, p. 47. http://www.ipcc.ch/pdf/assessment-report/ar4/syr/ar4_syr.pdf. Accessed 25 Feb 2016.
United Nations Framework Convention on Climate Change (UNFCCC). (2005). *Climate change small island developing states.* Bonn: UNFCC Secretariat. http://unfccc.int/resource/docs/publications/cc_sids.pdf. Accessed 25 Feb 2016.
United Nations General Assembly. (1948). Universal Declaration on Human Rights. http://ccnmtl.columbia.edu/projects/mmt/udhr/. Accessed 25 Feb 2016.
United Nations General Assembly. (2014). Draft outcome document of the third international conference on small island developing states {52.a}. http://www.sids2014.org/content/documents/358A-CONF-223-5%20ENGLISH.pdf. Accessed 25 Feb 2016.
United Nations High Commissioner for Refugees (UNHCR). (2007). Summary record of the 616the meeting of the Executive Committee of UNHCR (A/AC.96/SR.616), at para 26. http://www.unhcr.org/473c1c362.pdf. Accessed 25 Feb 2016.
United Nations High Commissioner for Refugees (UNHCR). (2008). Climate change, human rights and forced human displacement: Meeting report, Australia, 10 Dec 2008.
United Nations High Commissioner for Refugees Regional Office for Australia, New Zealand, Papua New Guinea, the Pacific. (2010). Pacific islanders face the reality of climate change, *Refugee Newsletter*, No.1/2010.
United Nations News Center. (2015). UN responding to 'devastating' impact of Tropical Cyclone Pam in Vanuatu, Pacific region, 14 March 2015. http://www.un.org/apps/news/story.asp?NewsID=50327#.VS1dXWY3M7A. Accessed 25 Feb 2016.
United Nations, Economic and Social Commission for Asia and the Pacific (ESCAP). (2000). Ministerial conference on environment and development in Asia and the Pacific 2000, Kitakyushu, Japan, 31 Aug–5 Sep 2000: Ministerial Declaration, Regional Action Programme (2001–2005) and Kitakyushu Initiative for a Clean Environment, Part Three. Sales No. E.01.II.F.12.
Von Doussa, J. (2008).Climate change and human rights. Centre for Policy Development. http://cpd.org.au/article/climate-change-and-human-rights. Accessed 25 Feb 2016.
Ward, G. (2000). Land tenure in the Pacific Islands: Changing patterns and implications for land acquisition. In R. Manning (Ed.), *Resettlement policy and practice in Southeast Asia and the pacific* (pp. 75–87). Manila: Asian Development Bank.
Warner, K., Afifi, T., Martin, S. F., Kälin, W., Leckie, S., Ferris, B., & Wrathall, D. (2013). *Changing climates, moving people: A framework for distinguishing and refining climate policies to address migration, human displacement and planned relocation* (Policy Brief, Vol. 8). Bonn: United Nations University Institute for Environment and Human Security (UNU-EHS).
Weir, T. & Virani, Z. (2010). *Three linked risks for development in the Pacific islands: Climate change, natural disasters and conflict.* PACE-SD Occasional Paper No. 2010/3.
Wilson, M. A., & Hunt, P. (Eds.). (2000). *Culture, rights and cultural rights: Perspectives from the South Pacific.* Aotearoa : Huia Pubishers.
Winch, P. (1964). Understanding a primitive society. *American Philosophical Quarterly.*
World International Property Organisation (WIPO). (2013). *Customary law, traditional knowledge and intellectual property: An outline of the issues.* http://www.wipo.int/export/sites/www/tk/en/resources/pdf/overview_customary_law.pdf

Worliczek, E., & Allenbach M. (2011). Customary land tenure and the management of climate change and internal migration. The example of Wallis Island. *Land Tenure Journal* (2) Thematic issue on land tenure and climate change. http://www.fao.org/nr/tenure/land-tenure-journal/index.php/LTJ/article/view/36. Accessed 25 Feb 2016.

Zahir, S., Sarker, R., & Al-Mahmud, Z. (2009). An interactive decision support system for implementing sustainable relocation strategies for adaptation to climate change: A multi-objective optimisation approach. *International Journal of Mathematics in Operational Research, 1*(3), 326–350.

Zbigniew, W., & Kundzewicz, L. J. M. (2007). Fresh water resources and their management. In *Climate change 2007: Impacts, adaptation and vulnerability, IPCC Fourth Assessment Report (AR4)* (p. 175). Geneva: IPCC.

Zorn, J. G. (2003). Custom then and now: The changing Melanesian family. In A. Jowitt & T. Newton Cain (Eds.), *Passage of change: Law society and governance in the pacific*.

Chapter 9
The Role of Remittances in Risk Management and Resilience in Tuvalu: Evidence and Potential Policy Responses

Sophia Kagan

9.1 Introduction

Although migration from low-income countries still draws negative connotations as a result of the perception of 'brain drain', the exponential growth of remittances is often credited with catapulting migration back into the development agenda (de Haas 2012).

However, while remittances may over time alleviate a household's poverty and hardship, this may be threatened by exposure to different shocks, whether economic, environmental, health or socio-political, which can push the household back into poverty. This issue is particularly critical for many countries in the Pacific that are experiencing significant adverse effects associated with climate change (Campbell 2010). Can remittances increase Pacific Islanders' ability to manage risk associated with environmental shocks, such as natural disaster or slow onset changes such as coastal erosion or flood surges? And if so, how can public policies facilitate or complement remittances to ensure that the most vulnerable households and communities are better able to prepare for and cope with environmental shocks?

In this chapter, the term remittances will be taken to mean both the funds and goods transfers from diasporas or migrant workers in one country, to their country 'of origin'. There are many of types of transfers that arguably take place between migrants and their communities of origin which will not be considered. For example, this discussion will not cover 'social remittances' that encompass ideas and behaviors that migrants bring back to their communities either through exchange of communication or through visits (by migrants to communities of origin or vice

---

S. Kagan (✉)
Technical Officer, International Labour Organization, Suva, Fiji
e-mail: sophiakagan@gmail.com

© Springer International Publishing Switzerland 2016
A. Milan et al. (eds.), *Migration, Risk Management and Climate Change: Evidence and Policy Responses*, Global Migration Issues 6,
DOI 10.1007/978-3-319-42922-9_9

versa) (Levitt and Lamba-Nives 2011; Levitt 2001). Although these exchange may have a very important relationship to risk management in communities of origin this is a separate area of analysis that deserves to be addressed.

This chapter considers the question of how remittances impact on risk management at the household level, in the context of environmental shocks in the small island country of Tuvalu, and outlines possible policy responses that could be taken to improve the positive effect of remittances. It is structured as follows: Section 9.2 will set out a conceptual framework for understanding risk management and will explore the theoretical relationships between remittances and different components of risk management. Section 9.3 will review the empirical evidence of remittance flows and uses of remittances to assess how this impacts on a household's ability to protect itself and cope with an environmental shock. Section 9.4 then looks at the case study of Tuvalu – one of the most environmentally vulnerable countries in the world, as well as one of the most reliant on remittances. The section will analyze existing secondary research on remittances to Tuvalu to better understand its nature and use. The chapter will conclude with policy recommendations for better understanding and supporting risk management through remittances in Tuvalu, which may also apply to countries in similar situations.

9.2 The Conceptual Relationship Between Remittances and Risk Management

The terms risk management, resilience, adaptive capacity and vulnerability have emerged in parallel out of different strands of academic and policy literature, including development discourse, scientific analysis (of natural disasters) and climate change adaptation research (Gaillard 2010). This sometimes makes the discussion of risk management in the context of environmental issues fraught with definitional confusion. This Section will apply a generic framework for conceptualizing risk management, which was adapted from Ehrlich and Becker (1972) in the 2014 World Development Report, *Risk and Opportunity: Managing Risk for Development*. It will then outline how remittances fit into this framework.

9.2.1 *A Conceptual Framework for Risk Management*

The 2014 World Development Report (WDR) defined risk management as 'resilience in the face of adverse events and prosperity through the pursuit of opportunities' (World Bank 2014a, 11) and set out a framework for risk management which looks both at the capacity to prepare for risk and the ability to cope afterwards.

The model defines four components of risk management: (i) **knowledge** of potential shocks and outcomes, (ii) **protection** measures to reduce probability and size of loss; (iii) **insurance** to transfer losses across people or over time; and (iv) **coping** to recover from losses.

The foundation for risk management is an understanding of what risks may occur and ways to reduce these risks, which is an important first step towards reducing households' and individuals' vulnerability. 'Knowledge', in this framework, comprises more than just accumulation of information but an appreciation of what that practically means for the individual or household and that allows them to make an assessment of their risk of exposure. It also encompasses the knowledge of what can be done to limit that exposure. For example, knowledge that cyclones may strike and destroy one's house is not the same as specific knowledge about cyclone risks to a particular coastal area combined with information about construction methods for more cyclone-resistant housing.

Based on the right knowledge, protection measures reduce risks by lowering the probability that negative impacts from shocks will occur. Using the earlier example, households that use information about cyclones can build cyclone-resistant housing, lessening the probability of damage and loss of life from such natural disasters.

To the extent that protection cannot completely eliminate the possibility of a negative outcome, insurance can help soften the impact from adverse shock by putting away or pooling funds that can be used in time of need. While some formal insurance systems might exist, often in developing countries individuals engage in 'self-insurance' such as holding savings or assets that can be sold in the event of a shock.

Coping is what occurs after a shock occurs. This can obviously involve using the resources that households obtained through preparation strategies such as insurance. However, it can also be unplanned coping mechanisms, such as selling assets to rebuild housing and livelihoods, increasing the amount of time spent working and rebuilding, migrating to another location, or taking out loans.

9.2.2 What Part Do Remittances Play?

Conceptually, there are several ways in which remittances may impact on a household's or community's ability to apply one or more of the four components of risk management discussed above. Most clearly households that are limited from managing their risk because of a lack of financial or other resources can benefit from remittances. For example, having additional funds can help households to purchase materials that can improve their resilience to environmental shocks (such as better construction materials, rainwater catchment tanks), or to set aside savings as 'self-insurance', to help them to deal with environmental shocks. Having access to extra funds from abroad can also be critically important as a coping strategy as it provides a source of income when the household's existing income stream has been destroyed (Scheffran et al. 2012).

Households may also be constrained not just because of lack of money but also lack of access to goods they need. This is particularly important for isolated communities as they often cannot get access to adaptive technologies at a good price. They are also most difficult to reach in the wake of disasters, particularly as government authorities are overwhelmed with delivering supplies across the country. Remittances in the form of sending or delivering goods may thus also be an effective strategy for improving risk management – whether it is supplying building materials for housing, or life-saving goods such as food or medicine after a disaster.

However, remittances may be less relevant to better risk management where the obstacle is not due to a lack of resources. For example, a household may not adequately prepare for an environmental shock because they do not have access to the right information to avoid risk, or lack the ability to understand or act on this information (World Bank 2014a). Alternatively, the household may appreciate the risk but fail to act on it due to a false sense of security, social norms or other behavioral factors.

9.3 Overview of Empirical Evidence

The relationship between remittances and risk management in the context of environmental shocks has been empirically studied in a number of countries, often looking at how remittances impact on coping strategies following disaster (Mohapatra et al. 2012).

It is interesting to note that research has generally focused on the individual resilience of households, rather than community resilience. However the inherent assumption that what is good for the individual receiving remittances is good for the community may not be borne out in practice. For example, Craven (2015) notes that while remittances in Vanuatu have strengthened the resilience of remittance-receiving households (for example, through construction of permanent housing) this has inadvertently increased vulnerability of the community due to the large scale erosion and clearing of traditional vegetation to make way for the increasing number of houses. In particular *"landslides have become an everyday reality for the community and, as more houses are built, this is likely to continue. Increased construction is also placing a stress on the community's finite water source…[due to] an exponential rise in the number of households installing Western-style showers and flushing toilets."* (Craven 2015: 6).

9.3.1 Remittances as a Coping Strategy

Studies since the mid-1980s have shown that migrants send more money as remittances after an environmental shock, such as a drought, flood or cyclone (Lucas and Stark 1985; Miller and Paulson 1999). Although many of the early studies looked at macroeconomic correlations between environmental shocks and volume of remittances to a region, more recent studies analyzed the relationship at the household level, testing whether the correlations are the same for all households or noticeably higher for poorer or needier households. For example, Funkhouser (1995) found that remittances to households in Nacaragua and El Salvador tend to be lower when the household head is working; while Clarke and Wallsten (2003) showed that households in Jamaica that faced the greatest damage during a hurricane received higher remittances than those facing less damage. This suggests that remittances do play a role in assisting households to cope with environmental shocks. However the evidence also shows that remittances do not necessarily restore households to their original position before the shock – in the case of Hurricane Grant in Jamaica, remittances only cover 25% of the losses due to hurricane damage (Clarke and Wallsten 2003; World Bank 2005).

Further research looking at remittances as a coping strategy was undertaken following the 2004 tsunami in the Indian Ocean. Many of the countries in Asia affected by the tsunami had already been strongly reliant on remittances. In Sri Lanka, following the devastation of the tsunami, research found that remittances could often be accessed more quickly than government relief funds. Although some remittance channels were affected by the destruction of infrastructure (such as banks) and loss of documents, remittances reached many communities when returned migrants hand-carried money home, though informal remittance-sending channels were also commonly used (Deshingakar and Aheeya 2006). Where remittances could be accessed, they played a critical role in filling gaps in government assistance, such as food not provided by the relief operation, medicines for chronic diseases, assistance to rebuild livelihoods and payment of interest on pre-tsunami debts (Deshingakar and Aheeya 2006; IOM 2007).

Similar studies have also been conducted in the Hindu Kush Himalayan region, indicating that remittances have often been used to procure food and other basic needs during or in aftermath of a disaster, and recreate livelihoods (Suleri and Savage 2006; Banerjee et al. 2011).

The most recent literature looking at remittances and environmental shocks is from 2012 Cyclone Evan in the Pacific, which particularly impacted on Samoa. In the aftermath of the damage, studies show that up 90% of disaster-affected households in Samoa received some form of international remittances, and of these, 72% received them within a week after the event (Le De et al. 2015). Despite the impact of the disaster on infrastructure, including telecommunications networks, 17.5% of those receiving remittances accessed remittances the same day of the tsunami, 24.5% within 1 and 3 days, and 30% between 3 days and 1 week after the event (Le De et al. 2015).

9.3.2 Remittances as a Strategy to Prepare for Risk

In general, far less analysis has looked into how remittances can be used as a strategy for reducing or managing risk, either through protection measures or insurance.

Mohapatra et al. (2012) cite a small number of studies which have linked remittances to construction of concrete housing which is more resistant to environmental impact than traditional mud houses, and greater access to communication equipment, suggesting that they are better prepared against natural disasters. For example, 77 % of Ghanaian households that received international remittances had concrete houses, in comparison with 68 % of comparable households that did not receive remittances.

Another important measure for improved protection is access to electricity and communication facilities such as fixed or mobile telephones, which can help households to receive warnings and communicate with others in the situation of an emergency. Mohapatra et al. (2012) show that 80 % of households that receive remittances from OECD countries have electricity in comparison to 69 % of comparable houses that do not receive remittances. Similarly there was also a (smaller) distinction in access to telephones with 69 % of households with remittances from OECD countries having a mobile phone, while only 55 % of comparable households without remittances had one.

Some similar results were found in Burkina Faso using the results of a national household survey, which found that 30 % of households receiving remittances had concrete houses, in comparison to 25 % of comparable houses without access to remittances (Mohapatra et al. 2012). Another region where remittances have been found to have an impact on housing quality is in the Himalayan areas, where remittances have been used for disaster preparedness such as strengthening of housing quality or procurement of boats in flood-affected communities and buying irrigation equipment in settlements likely to be affected by drought (Banerjee et al. 2011).

Qualitative evidence of the role of remittances in risk management is also limited. A review of the engagement of diasporas from Asian countries affected by the 2012 tsunami by the International Organization for Migration presented significant evidence of *ex-post* support (in the form of mobilizing donations), but cited only a few examples of *ex-ante* assistance. One example given was an Indian diaspora organization which sponsored a visit by a qualified member of the diaspora to facilitate the development of a national Indian disaster mitigation plan in response to the 1991 Gujarat earthquake. The organization further planned to develop a disaster management resource center to mitigate and respond to future disasters (IOM 2007).

9.4 Case Study of Tuvalu

Tuvalu is one of the world's smallest independent states in terms of land area (26 km^2) and population (11,206) (Tuvalu Government 2012). Also recognized as one of the most environmentally vulnerable states in the region, it has been classified as a Least Developed Country by the United Nations almost since its independence in 1978. Entrenched limitations in stimulating private sector growth have limited employment opportunities and resulted in an extensive history of labor migration overseas, including to Nauru's phosphate mines, international seafaring and agricultural work in New Zealand.

In 2012 the population census recorded an economically active population of 4243, with 51% classified as 'paid workers', however unemployment and underemployment are known to be significant challenges (Tuvalu Government 2012).[1] In particular, a serious limitation on employment growth is due to the tiny private sector (government remains the largest employer in Tuvalu, employing two-thirds of paid workers).

Tuvalu is heavily dependent on aid and remittances with more than 40% of the population relying on remittances (Tuvalu Government 2012). Although the government takes a positive attitude to remittances – particularly from temporary migrants such as seafarers and seasonal workers (described in further detail below) – there is no government policy on the use of remittances for risk management and improved resilience to climate change. Indeed remittances and migration are generally kept separate from the issue of climate change, due to the Tuvaluan government's sensitivity to the issue of 'climate change induced migration'. Senior members of government have repeatedly indicated that mass migration will not be an adaptation strategy except in the case of a tsunami or similar disaster event (Smith and McNamara 2015) in order to ensure that Tuvalu retains its sovereignty. It also enables the government to better assert pressure on other nations to commit to lower emissions targets (PACNEWS 2015).

9.4.1 Climate Change Impact and Risk Management

According to the International Panel on Climate Change, Tuvalu is at extreme high risk of a number of climate change impacts and associated disasters (IPCC 2007). Some of the likely impacts include health, economic and social consequences, including (Tuvalu Government 2011):

- intense storms and tropical cyclones and associated damages to livelihoods, infrastructures;
- droughts and associated health and economic impacts;
- extreme rainfall and associated flooding and health implications;
- sea level rise and increased erosion and inundation, impacts on marine systems; and
- increase in temperature and consequent health impacts, as well as impact on marine systems due to coral bleaching and fish stocks.

---

[1] Problems of un- and under-employment have also been exacerbated by continued urbanization of the country's capital, Funafuti, as those in the working age populations move from the outer islands to the main urban center in search of employment. By 2012 over half (57%) of Tuvalu's total population of 10,782 was living on Funafuti, compared with only 15% of the population of 5,887 in 1973 (Tuvalu Government 2012).

Table 9.1 Adaptive responses to environmental stressors

Type of adaptive response (% of respondents that used this response)	Example(s) of adaptive response
Bear the effects (53 %)	Do nothing, accept loss, and/or put faith in God to deal with the financial burden of cultural commitments, loss of land due to coastal erosion and the loss of crops due to salt water intrusion
Reactive short-term behavior change (79 %)	Change pattern of water use (ration and conserve water, use seawater) and demand (slaughter pigs and abandon crops during periods of water shortage
Engage in the market (66 %)	Purchase water from the government during water shortages, purchase food from the store to cope with limited availability of local food, get an education to secure a job to make more money in response to high cost of living
Rely on local resources and knowledge (45 %)	Fish to deal with the high cost of store food
Share the burden (53 %)	Share food among family members (especially local food from the outer islands)
Rely on aid or government intervention (71 %)	Receive water tanks from international donors to cope with dry spells
Change locations to reduce exposure (24 %)	Evacuate homes prone to inundation or migrate overseas in response to poverty of opportunity in Tuvalu
Modify surroundings to reduce exposure (34 %)	Construct coastal barriers to reduce effects of inundation and erosion, elevate crops to deal with saltwater intrusion

Source: Adapted from McCubbin et al. (2015)

These environmental issues, however, are often not perceived as standalone issues but instead contribute to a vulnerability that 'is deeply interwoven in the historical, socioeconomic, and cultural fabric of communities' (McCubbin et al. 2015). McCubbin et al. found, through interviews with participants in Funafuti, that while a number of environmental factors were identified as stressors on livelihoods, including food security, water, sea level rise, strong winds and temperature, they were often closely interlinked with other perceived issues in terms of over-crowding and a lack of job opportunities.

Studies of adaptive responses to climate change in Tuvalu are limited. While there is a multitude of climate change development projects in Tuvalu funded through overseas aid (such as the EU-funded Pacific Climate Change Alliance) there is less evidence of community or individual-led 'bottom up' capacity building. McCubbin et al. (2015), when probing the adaptive responses of participants in Funafuti found that amongst the proactive strategies to deal with impacts, reactive, short-term behavior was most common and more long-term strategies such as modifying surroundings to reduce exposure and changing locations to reduce exposure were utilized far less frequently (Table 9.1).

Delving into the reasons why more durable adaptation measures are not undertaken is a challenging exercise, as the decision-making processes are complex and nuanced. McCubbin et al. found that in the context of access to water, a common constraint to adaptive capacity was financial – households had insufficient funds to purchase additional tanks. However, another reason proffered had to do with donor interventions that had introduced perverse incentives to wait for external assistance, rather than taking the risk by spending money on a tank. A similar confluence of reasons was found in relation to actions responding to coastal erosion. While some households had modified their surroundings (e.g. by constructing sea walls or elevating gardens) or changed location (e.g. by shifting their house or relocating their crops) to reduce their exposure, these cases were not common. Partly this was again found to be due to a lack of financial resources to make changes, as people who occupy the most exposed areas tended to be the ones who come from outer islands with fewer job prospects, and often more resource-constrained (Tuvalu Red Cross 2012). Additionally religious convictions were also articulated, as well as the belief that because coastal inundation had become prominent in the international arena, the international community would assist them to cope, thus obviating the need for their own adaptive strategies.

9.4.2 Remittances and Migration

Tuvalu has an extensive history of labour migration overseas dating back to the late 1800s (while still a part of the Gilbert Islands). Temporary migration for economic gain because particularly important in the early twentieth century when Tuvaluan men migrated to Banaba and Nauru for phosphate mining. However the depletion of resources in Banaba in 1979, and in Nauru in the late 1990s forced all Tuvaluan workers to eventually return home (though some did manage to move to New Zealand).

Another longstanding opportunity for temporary migration for Tuvaluan men is onboard foreign merchant vessels as seafarers. Seafaring has also had a long history and since independence in 1978, Tuvalu has had a Tuvalu Maritime Training Institute (TMTI) (formerly Tuvalu Maritime School) which has remained the anchor of the country's engagement with international labour migration. In early 2006 there were up to 440 Tuvaluan seafarers, comprising nearly 15 % of the adult male population (Asian Development Bank 2007). However, numbers of workers have dropped dramatically since the Global Financial Crisis, and now stand at just over 100 per year.

Since the late 1980s, Tuvalu has also participated in temporary employment programs to New Zealand. The New Zealand/Tuvalu Guest-Worker scheme was introduced in 1986 and often up to 80 workers the chance to be employed for a up to 3 years. The Guest-Worker program was phased out in the 1990s and was replaced by a temporary seasonal worker program in 2007. This scheme brings seasonal workers from several Pacific Island countries to work in horticulture and viticulture and a similar scheme has operated in Australia from 2008. However the number of workers from Tuvalu has been low – with less than 30 requested by Australian

Table 9.2 Percentages of households receiving remittances, 1979, 2002 and 2012

Island	1979	2002	2012
Nanumea	63	51	47
Nanumanga	64	78	37
Niutao	68	53	47
Nui	57	62	45
Vaitupu	54	34	43
Nukufetau	81	61	48
Funafuti	*24*	*41*	*36*
Nukulaelae	88	37	31
Niulakita	11	0	57
Tuvalu	54	47	40

Sources: Tuvalu Government (2015)

approved employers, and 78 by New Zealand workers in the 2013–2014 financial year (Hay and Howes 2012).

Permanent migration, particularly to New Zealand, has offered another migration avenue. Both Australia and New Zealand operate skilled migration schemes which allocate points for qualifications and other characteristics of migrants. Thus far, the majority of these workers have migrated to New Zealand, though the number has been low – approximately 1432 between 2002 and 2011 (Curtain and Powell 2011).

Another permanent migration option, available only through the New Zealand government is the Pacific Access Category (PAC) which was initiated in 2002. The scheme grants a permanent resident permit but is subject to a quota of 75 people per year for Tuvalu and Kiribati respectively (including applicants and dependents), and 250 people for Tonga and Fiji. Applicants must fulfil certain educational, English fluency and residency in Tuvalu requirements and will only be granted a visa if on the receipt of a job offer in New Zealand, offering a minimum designated income. Between 2002 and 2011 around 908 people gained entry to New Zealand via the PAC (Curtain and Powell 2011).

With prospects for nationally driven economic growth, Tuvalu is heavily dependent on aid and remittances. Currently, 40% of households in Tuvalu receive remittances though this has decreased since 1979 when 54% of households received remittances (Tuvalu Government 2015) (Table 9.2).

Remittances are particularly important to outer islands where there are few employment opportunities. For example, on the island of Nanumea remittances grew from half the island's income in the 1970s and 1980s to 75% in the 1990s in large part because of the collapse of copra marketing as world prices slumped (Chambers and Chambers 2001).

The most significant source of remittances are seafarers. This is due to the relatively long duration of contracts (1 year), the low cost of living on ships, and the fact that transport is paid by the employer rather than the worker, contributing to higher take-home wages. Data from 2004 to 2005 shows that remittances from seafarers and relatives living overseas was the fourth largest source of household income, at 9 % of total income (Tuvalu Government 2006). A large share of seafarer earnings return directly to Tuvalu, due to limited options and needs for expenditure aboard ships (lodging and board are provided), and limited port time during which wages can be spent. Seafarers remit 70–90 % of their base wage directly back home to Tuvalu through manning agency accounts with the National Bank. The monthly average in remittances from seafarers in 2002 was near to AUD $452,000 but by 2007 the monthly amount remitted had dropped to $209,000. In 2008, with the onset of the global financial crisis, the monthly amount fell to a low of only $68,000 per month (World Bank 2015).

Remittances from seasonal workers are generally lower than for seafarers, but still an important income generating strategy for many households. Unlike seafaring, accommodation costs can be relatively high in Australia and New Zealand, and seasonal workers are also required to pay a proportion of the airfares to bring them to the destination country. The Tuvaluan government does not collect specific information on seasonal workers' wages, earnings or remittances. However, other studies undertaken by the World Bank and others suggest that over a 6 month employment period at 40 h per week, gross earnings would amount to NZ$ 13 152 for the contract period, and net earnings (after deduction of tax and living expenses) are estimated at NZ$ 3718 (US$ 2820), or NZ$ 620 (US$ 470) per month of work. This is in contrast to other calculations which estimate that after deducting all costs (travel, accommodation, food), on average workers earned NZ $5,764 per season (McLellan 2008).

The final channel of remittances which will be explored is remittances from the New Zealand 'diaspora', or in other words, persons of Tuvaluan ethnicity, living in New Zealand. The 2013 Census recorded 3459 people of Tuvaluan ethnicity residing in New Zealand (not including seasonal workers), with the majority living in Auckland. Of the ethnic Tuvaluans in New Zealand, around 39 % were born in Tuvalu, suggesting relatively recent migration.

The median income for those aged 15 and over was identified to be NZ$ 14,600, however the overall median disguises the large discrepancy between men and women, as men's median income of NZ $20,700 was more than double that of Tuvaluan women in New Zealand (NZ $9900). Both were significantly below the median incomes for other Pacific ethnicities, and the New Zealand population as a whole. Tuvaluans were most commonly employed, for men this was predominantly manufacturing (17.3 %) while women were most likely to work in health care and social assistance (23.9 %).

While remittance data is not easy to obtain, Boland and Dollery (2007) have reported remittances from New Zealand for 2003 to be less than A $400,000. Boland and Dollery have also reported non-commercial imports from New Zealand from Tuvalu Department of Customs and Taxation (Boland and Dollery 2007) as a proxy for gifts and purchases for use in Tuvalu. For 2003 these stood at around A$ 190,000. These amounts are small compared to remittances from seafarers (Simati and Gibson 2001).

9.4.3 Use of Remittances

Evidence on the use of remittances to Tuvalu is limited as surveys asking about remittances generally ask about the volume of remittances but not information on spending. The evidence that is available from smaller research surveys suggests that most remittances are used for food, children's education and community feasts for occasions such as weddings or funerals (Simati 2009). This is consistent with survey data from other Pacific Island countries (Connell and Brown 2005).

Using survey data from communities in Auckland, Simati (2009) shows that remittances from New Zealand do not flow continuously or on a frequent basis, but are more likely to occur in response to special request, for example for weddings or funeral expenses (Simati 2009). Simati notes that 'with regards to the migrants' contributions to the families remaining in Tuvalu, overall the migrants agreed that a request, financial or otherwise, from families in Tuvalu must always be fulfilled.' One way in which this is sometimes done was to pool resources to respond to a request jointly with each person contributing a small amount (Simati 2009).

Research from seafarers suggests that remittances may be commonly used for housing construction, but also purchasing small luxury goods. As remittances from seafarers tend to be higher in volume and more regularly sent from remittances from diaspora abroad, there may be greater scope for investment in such things as better housing. A study of seafarers from Kiribati and Tuvalu also indicates that some money is saved by the seafarer himself – with data showing that up to 37 % of remittances from seafarers were transferred into a personal bank account rather than to the seafarer's family's account (Borovnik 2006). Some funds are also transmitted to the National Provident Fund as voluntary contributions.

Another common channel for remittances is through community groups and churches. As most Tuvaluans in New Zealand belong to the same religious denomination, church activities and networks are a very important part of Tuvaluans living abroad and an opportunity to meet. Tuvaluans in New Zealand are generally required to contribute annual to the church's headquarters in Tuvalu (Simati 2009).

Also common are contributions to the migrants' home island, and community development projects that are undertaken by the local government (Connell and Brown 2005), however much more research is needed about precisely the types of activities that are undertaken. One example is *Kautaina Nanumea*, which is a loose organization of people in Nanunumea (an island in Tuvalu) or living abroad in Fiji or New Zealand. *Kautaina Nanumea* organizes a biennial Nanumea Development Forum to discuss development issues on the island, and collects remittances from overseas-based Nanumeans for community project – such as improving the school system on the island.

An interesting aspect of remittances in Tuvalu, which is consistent with data on remittances to other Pacific Island countries, is the level of distribution and sharing of remittances – which in many cases reach the most vulnerable (McCubbin et al. 2015). For the most part however, the evidence is largely qualitative and based on small participant size (Simati and Gibson 2012). Evidence from the Household

Income and Expenditure Survey suggests that most households participate in giving and receiving gifts but that those most in need are likely to receive more than they give (World Bank 2014b), using a sick household member as a proxy. More concrete evidence exists in Fiji and Tonga showing that households below the poverty line can expect to receive more remittances, the poorer they are or they become but this study cannot necessarily be extrapolated to Tuvalu (Brown et al. 2014).

In terms of other uses for remittances, there is so far little evidence of funds being used towards business entrepreneurship. With the exception of a small number of returned seafarers that started a mechanics business, few families have been able to start small businesses with the remittances that they receive from abroad, though this needs to be subject to further research. Anecdotal evidence suggests that the reason is to do with credit constraint, a lack of entrepreneurship training, and a culture of sharing that disincentivizes individual risk taking as other members of the family can request for profits.[2]

Overall, there is little evidence to suggest that remittances have been used explicitly for protection or insurance to prepare for, or cope with, environmental shocks, with the exception of construction of more permanent housing. Of course personal savings and contributions to the Provident Fund, which is more common amongst families with seafarers, can be seen as a form of insurance (money can be temporarily taken from the Provident Fund, but an interest rate is applied). Once again, the reasons are complex and need to be explored in more detail. However the next section provides some areas for consideration, which may help to better understand, and strengthen, the relationship between remittances and risk management.

9.5 Policy Implications and Possible Responses for Tuvalu's Government

The Tuvaluan government has been a strong advocate of increasing migration opportunities for its people, particularly as seafarers and seasonal workers. As remittances do play an important role in supporting households in basic consumption, particularly on the outer islands where there are few income-generating opportunities, this is an important foundation for a discussion of risk management and further policy responses.

9.5.1 Current Policy Framework on Remittances

The key documents relating to climate change and mobility in Tuvalu are the Tuvalu National Strategic Action Plan for Climate Change and Disaster Risk Management (2012) and the National Labor Migration Policy (2015).

---

[2] Interviews with local community members (February 2014).

The Tuvalu National Strategic Action Plan does not focus explicitly on remittances, but does include a section on increasing migration and establishing professional training programs in key identified occupations to allow for employment in neighboring countries if climate change migration is necessary. The policy looks not only at voluntary migration, but also forced migration and relocation due to impacts of climate change including sea level rise. From an institutional perspective, there has been little progress made towards any of the above activities until recently. Partly, this may be due to the fact that the Action Plan was developed under a former Tuvalu Prime Minister, Willy Telavi, who was succeeded by Enele Sopoaga in 2013. Soon after his appointment, Sopoaga indicated disagreement with key aspects of the Strategic Action Plan, particularly noting that that relocating Tuvaluans to avoid the impact of sea level rise "should never be an option because it is self-defeating in itself. For Tuvalu I think we really need to mobilize public opinion in the Pacific as well as in the [rest of] world to really talk to their lawmakers to please have some sort of moral obligation and things like that to do the right thing" (Radio New Zealand, 3 September 2013).

The National Labor Migration Policy, endorsed by Cabinet in August 2015, does contain more detail on the government's policy relating to remittances, outlining their importance as a component of household incomes for many households in Tuvalu. The Action planned appended to the Policy focuses on a two-pronged goal for remittances including – 'find[ing] ways to harness remittances for productive development at home, as well as ensuring that the costs of sending remittances are minimized'. These activities are expected to be undertaken by government between 2016 and 2019.

9.5.2 Further Initiatives Which Could Be Explored

In terms of helping households to better plan for use of remittances, several initiatives could be explored, including providing opportunities for micro-insurance, and increasing the financial literacy of individual migrants and their family members to enable them to make informed decisions regarding how to save or use remittances, including mechanisms that could improve the households' or communities' resilience to climate change. As many decisions in Tuvalu are made at a community level, engaging with community and other organizations is important.

Another reason to consider engaging with community organizations is that these can be an important conduit for distributing remittances especially vulnerable individuals in communities. The increasing engagement of churches in climate change campaigns can provide a positive avenue for remittances to be used for increased risk management (for example, recent research and engagement by the Pacific Conference of Churches in the issue of climate change displacement). By helping to ensure that church and community groups are more informed, empowered and trained in the area of risk management, they can also help to inform their community members, and better plan for use of remittances in a way which improves the communities' ability to cope. It may also counteract the passive attitude that donor infrastructure has created in some communities.

Engagement with the Tuvaluan community abroad could also be further pursued in order to help direct remittances to activities that are likely to achieve the greatest resilience. For example, the government may consider communicating potential gaps or constraints in government policies on resilience that could be met through remittance contributions. Other Pacific countries (especially Tonga and Samoa) have made extensive use of their New Zealand and Australia-based diaspora including exploring options to make remittance sending cheaper and quicker – particularly in the wake of a disaster.

Finally, there is an urgent need to better understand the needs of the most vulnerable communities, including those without access to land, without migrant relatives or other vulnerabilities such as disability, and to determine whether remittances and other forms of exchange are reaching them. Following this research, government can play a more active and targeted role in providing some level of basic support so that such individuals are afforded some level of protection.

In conclusion, this chapter has explored the relationship between remittances and risk management in the context of environmental shocks. It has outlined the role that remittances can play in bolstering ex-ante preparation and ex-post coping strategies. Whilst evidence of remittances is much stronger in relation to improving households' ability to cope, there is also support for the link between remittances and more climate resistant housing, better access to communication and information. In Tuvalu, remittances are a critical part of coping strategies to environmental changes, providing access to food and education and in some cases, more durable housing. However, much more work is required by government to ensure that remittances can more effectively be used to strengthen risk management.

References

Asian Development Bank. (2007). *Tuvalu 2006 Economic Report, from plan to action*. Available at: http://www.adb.org/sites/default/files/publication/28755/2006-economic-report.pdf. Accessed 20 Aug 2016.

Banerjee, S., Gerlitz, J. Y., & Hoermann, B. (2011). *Labour migration as a response strategy to water hazards in the Hindu Kush-Himalayas*. Kathmandu: International Centre for Integrated Mountain Development.

Boland, S., & Dollery, B. (2007, March). The economic significance of migration and remittances in Tuvalu. *Pacific Economic Bulletin, 22*(1). Asia Pacific Press.

Borovnik, M. (2006). Working overseas: Seafarers' remittances and their distribution in Kiribati. *Asia Pacific Viewpoint, 47*(1), 151–161.

Brown, R., Connell, J., & Jimenez-Soto, E. (2014). Migrants' remittances, poverty and social protection in the South Pacific: Fiji and Tonga. *Population, Space and Place, 20*(5), 434–454.

Campbell, J. (2010). Climate change and population movement in pacific island countries. In B. Burson (Ed.), *Climate change and migration: South Pacific perspectives*. Wellington: Institute of Policy Studies.

Chambers, K., & Chambers, A. (2001). *Unity of heart, culture and change in a Polynesian Atoll Society*. Prospect Heights: Waveland Press.

Clarke, G., & Wallsten, S., (2003, January). *Do remittances act like insurance? Evidence from a natural disaster in Jamaica.* World Bank Development Research Group.

Connell, J., & Brown, R. (2005). *Remittances in the pacific: An overview.* Asia Development Bank, Pacific Studies Series.

Craven, L. K. (2015). Migration-affected change and vulnerability in rural Vanuatu. *Asia Pacific Viewpoint.* doi:10.1111/apv.12066.

Curtain, R., & Powell, R. (2011). Background analysis and recommendations on designing demand-based technical and vocational educational and training frameworks for the pacific. *ADB Technical Assistance, 12*, 67–77.

De Haas, H. (2012). The migration and development pendulum: A critical view on research and policy. *International Migration, 50*(3), 8–25.

Deshingakar, P., & Aheeya, M. (2006). *Remittances in crisis: Sri Lanka after the tsunami.* Overseas Development Institute Humanitarian Policy Group Background Paper (August 2006).

Ehrlich, I., & Becker, G. (1972). Market insurance, self-insurance, and self-protection. *Journal of Political Economy, 80*(4), 623–648.

Funkhouser, E. (1995). Remittances from international migration: A comparison of El Salvador and Nicaragua. *Review of Economics and Statistics, 77*(1), 137–146.

Gaillard, J. C. (2010). Vulnerability, capacity and resilience: Perspectives for climate change and development policy. *Journal of International Development, 22*, 218–232.

Hay, D., & Howes, S. (2012). *Australia's pacific seasonal worker pilot scheme: Why has take-up been so low?* (Development Policy Centre Discussion Paper 17). Canberra: Crawford School of Economics and Government, The Australian National University. https://devpolicy.anu.edu.au/pdf/papers/DP_17_-_Australia%27s_Pacific_Seasonal_Worker_Pilot_Scheme.pdf. Accessed 29 Apr 2015.

International Organization for Migration. (2007). *Migration, development and natural disasters: Insights from the Indian Ocean Tsunami.*

Le De, L., Gaillard, J. C., & Smith, M. (2015). Remittances in the face of disasters: A case study of rural Samoa. *Environement, Development and Sustainability, 17*, 653.

Levitt, P. (2001). *The transnational villagers.* Berkeley: University of California Press.

Levitt, P., & Lamba-Nives, D. (2011). Social remittances revisited. *Journal of Ethnic and Migration Studies, 37*(1), 1–22.

Lucas, R., & Stark, O. (1985). Motivations to remit: Evidence from Botswana. *The Journal of Political Economic, 93*(5), 901.

McLellan, N. (2008). *Workers for all seasons? Issues from New Zealand's Recognised Seasonal Employer (RSE) program.* Hawthorn: Institute for Social Research, Swinburne University of Technology. http://www.sisr.net/publications/0805maclellan_RSE.pdf. Accessed 20 Aug 2016.

McCubbin, S., Smit, B., & Pearce, T. (2015). Where does climate fit? Vulnerability to climate change in the context of multiple stressors in Funafuti, Tuvalu. *Global Environmental Change, 37*(1), 43–55.

Miller, D., & Paulson, A. (1999). *Informal insurance and moral hazard: Gambling and remittances in Thailand.* Evanston: Mimeo, Kellogg Graduate School of Management, Northwestern University.

Mohapatra, S., Joseph, S., & Ratha, D. (2012). Remittances and natural disasters: Ex-post response and contribution to ex-ante preparedness. *Environment, Development and Sustainability, 14*(3), 365–387.

PACNEWS. (2015, May 11). Relocation to bigger countries is not an option, says Prime Minister Sopoaga. *PACNEWS*, http://www.pina.com.fj/?p=pacnews&m=read&o=1443256633555149414b5f02b8c9cc. Accessed 15 Aug 2015.

Scheffran, J., et al. (2012). Migration as a contribution to resilience and innovation in climate adaptation: Social networks and co-development in Northwest Africa. *Applied Geography, 33*, 119–127.

Simati, A. (2009). *The effect of migration on development in Tuvalu: A case study of Tuvaluan migrants and their families.* Thesis for Master of Philosophy in development studies. Massey University. New Zealand.

Simati, A., & Gibson, J. (2001). Do remittances decay? Evidence from Tuvaluan migrants in New Zealand. *Pacific Economic Bulletin, 16*(1), 55.

Simati, A. M., & Gibson, J. (2012). *Do remittances decay? Evidence from Tuvaluan migrants in New Zealand.* http://catalogue.nla.gov.au/Record/329499. Accessed 15 Apr 2015.

Smith, R., & McNamara, K. (2015). Future migrations from Tuvalu and Kiribati: Exploring government, civil society and donor perceptions. *Climate and Development, 7*(1), 47–59.

Suleri, A. Q., & Savage, K. (2006). *Remittances in crises: A case study from Pakistan* (An Humanitarian Policy Group background paper). London: Overseas Development Institute.

Tuvalu Government. (2006). Household income and expenditure survey (2004/2005).

Tuvalu Government. (2011). National climate change policy 2012–2020.

Tuvalu Government. (2012). Population & housing census, preliminary analytical report. Government of Tuvalu.

Tuvalu Government. (2015). *National Labor Migration Policy.* Available at http://www.ilo.org/wcmsp5/groups/public/---asia/---ro-bangkok/---ilo-suva/documents/publication/wcms_431831.pdf. Accessed 20 Aug 2016.

Tuvalu Red Cross. (2012). *TeKavatoetoe* community report. Tuvalu Red Cross.

World Bank. (2005). *Global economic prospects 2006: Economic implications of remittances and migration.* Washington, DC: World Bank.

World Bank. (2014a). *World Development Report 2014: Risk and opportunity.* Available at http://siteresources.worldbank.org/EXTNWDR2013/Resources/8258024-1352909193861/8936935-1356011448215/8986901-1380046989056/WDR-2014_Complete_Report.pdf. Accessed 10 Aug 2015.

World Bank. (2014b). *Hardship and vulnerability in the Pacific Island countries: A regional companion to the world development report 2014.*

World Bank. (2015). *Optimizing development benefits from international labour migration in Tuvalu* (not publicly released).

Part IV
Policy

Chapter 10
Remittances for Adaptation: An 'Alternative Source' of International Climate Finance?

Barbara Bendandi and Pieter Pauw

10.1 Introduction: Remittances and Adaptation Finance

Even the most stringent efforts to reduce greenhouse gas emissions cannot prevent climate change impacts in the next few decades, making adaptation essential (Klein 2010). Developing countries are historically least responsible for the emissions that result in climate change, but most exposed to its impacts. Those most vulnerable to climate change will be the poorest people in migration-prone areas of developing countries (e.g. Ayers 2011). The costs of adaptation in developing countries are difficult to assess, but were recently estimated in the order of hundreds of billions of US Dollars per year (UNEP 2014). Explicit international funding possibilities for adaptation activities however remain limited in scale. The 2009 Copenhagen Accord of the United Nations Framework Convention on Climate Change (UNFCCC) recognized that substantially greater financial resources are needed to support mitigation and adaptation in developing countries. In this Accord and the subsequent Cancun Agreements, developed countries pledge to mobilize USD 100 billion per year for this purpose from 2020 onwards, coming from *'a wide variety of sources,*

B. Bendandi (✉)
Ca' Foscari University, Venice, Italy

Policy Officer. United Nations Convention to Combat Desertification (UNCCD)
e-mail: 955970@stud.unive.it

P. Pauw
German Development Institute/Deutsches Institut für Entwicklungspolitik (DIE), Bonn, Germany

Utrecht University, Utrecht, Netherlands

Stockholm Environment Institute, Stockholm, Sweden
e-mail: pieter.pauw@die-gdi.de

public and private, bilateral and multilateral, including alternative sources of finance' (UNFCCC 2010; §8).

The sources of adaptation finance are not well understood. And to the extent that they can be tracked, they do not seem to mobilize the billions of adaptation finance that are needed. Concerning public sources, for example, the Adaptation Fund is often considered to be progressive and innovative. Yet the predictability and sustainability of its future funding are uncertain as it partly depends on the development of the Clean Development Mechanism's market (Horstmann and Chandani 2011; 435). Its future had to be safeguarded through a public capital injection during COP19 in Warsaw. Developed and developing countries have now pledged financial resources for the newly established Green Climate Fund, which aims to spend 50 % of its resources on adaptation, but its project pipeline still needs to be developed. Multilateral and Bilateral Development Banks are increasingly investing in adaptation, but the expenditure remains low compared to mitigation. The discussion on private sources of adaptation finance, or on private engagement in adaptation in general, is in its early stages (Pauw 2014). It remains hard to even identify public-private adaptation projects, let alone study the effectiveness, replication or upscaling potential of public-private adaptation interventions (c.f. Kato et al. 2014). Indeed, private financing for adaptation is difficult to track and seems minimal compared to private financing of mitigation (Buchner et al. 2012). What exactly is meant with the third 'alternative source' of climate finance has not been clarified.

This chapter brings together literature on climate finance and remittances–money sent to families and friends in the origin countries by migrants – and analyses whether remittances could be considered as an 'alternative' source of adaptation finance in international climate negotiations. An alternative source means it is neither disbursed by the public sector, nor can it be labelled as 'private finance' as there is no objective of having *'reasonable, relatively quick and predictable returns, at acceptable risks'* (see Pauw and Pegels 2013; 2).

Given remittances' increasing magnitude and potential to contribute to development, governments have already been employing policy measures to harness the remittance potential for investments with a long-term perspective (Aparicio and Meseguer 2012). Some literature shows that households that receive this type of support have also proven to be more resilient to external stressors including natural disasters (Yang 2008; Mohapatra et al. 2012; Ebeke and Combes 2013).

Migrant investors are distinguished from the traditional private sector because determinants for remitting might go beyond profit making and rates of return. Key drivers for investing in areas of origin include family bonds and networks, and thus altruism, prestige, implicit co-insurance agreements and perspectives of return (Straubhaar and Vadean 2006). The 'tempered altruism' or 'enlightened self-interest' that often drive remittance behaviour (Lucas and Stark 1985) makes diaspora investments particularly suitable for adaptation projects. The fundamental difference between individuals or groups either referred to as 'migrants' or 'the diaspora' lays in the willingness of the act. While migration is voluntary, diaspora is forced, either by physical or economic factors. Moreover, one of the key characteristics of diaspora is summarized by the 'leaving home and staying in touch' attitude

(CheSuh-Njwi 2015). Throughout this chapter we will refer to the concept of diaspora for the importance of the need to move away from the places of origin and the links maintained with the family members or the ancestral community.

The need for adaptation investments is often concentrated in the water and agriculture sectors, as the livelihoods of most of the people in developing countries depend on these sectors. However, compared to the large investments in energy and transport infrastructures required for mitigation, land-based sectors are far less attractive to 'traditional' private investors, particularly if they are in exposed disaster-prone areas. The motivation to finance adaptation thus often needs other drivers than monetary returns.

In this context, the potential for remittances to play a role as an 'alternative source' of adaptation finance analysed for the following reasons: (1) the recorded volume of these flows to developing countries -expected to raise up to USD 516 billion in 2016 by the World Bank- has tripled ODA since 2013, which was USD 134.8 billion (OECD 2014); (2) the direct connection with the household level often hard to be reached by public interventions; and (3) the motivation to remit, not only based on returns in profit but also on personal bonds, increasing the likelihood for remittances to be spent in remote areas, where the traditional private sector would not necessary invest and where need for adaptation measures might be higher.

This is, however, not enough to affirm that remittances could be an alternative source of adaptation finance contributing to the annual USD 100 billion pledge of developed countries. To identify whether remittances meet the UNFCCC's expectations of adaptation finance for developing countries, this chapter builds on ten climate finance criteria from the Copenhagen Accord and the Cancun Agreements as distilled by Pauw et al. (2015) and examines literature and existing empirical data on remittances against these criteria.

This chapter is structured as follows. The next section identifies the ten criteria for adaptation finance and a reference framework towards which recurring features of remittances will be analyzed. By applying these criteria, section three then reviews key findings on the remittances and considers the motivation to remit and the key drivers that might lead to adaptation finance initiatives at individual, household and community level. Section 10.4 will analyze remittances as flows and, as such, their potential for being leveraged as investments in adaptation. Section 10.5 will discuss the role of public institutions in guaranteeing appropriate frameworks for remittances to be channeled in a 'transparent' and 'balanced' way towards adaptation actions.

10.2 Adaptation Finance Criteria

This section builds on ten criteria for adaptation finance that were identified and defined by Pauw et al. (2015). They were elaborated for the purpose of this study, as provided in Table 10.1, which (i) lists the ten criteria that were identified for adaptation finance (predictable; sustainable; scaled up; provided with improved access;

Table 10.1 Ten climate finance criteria as distilled from the Copenhagen Accord and the Cancun Agreements (first and second column) as well as our interpretation of these criteria in order to analyse whether remittance can meet these criteria

Copenhagen accord	Cancun agreements UNFCCC	Interpretation to analyse remittances
Predictable (…) financial resources (…) to support the implementation of adaptation action in developing countries (§3)	Decision: (…), **predictable** (…) funding shall be provided to developing country parties (§97)	Can recipients anticipate these flows and thereby be able to react and plan accordingly to their adaptation needs?
Predictable (…) funding (…) shall be provided to developing countries (§8)		
Sustainable financial resources (…) to support the implementation of adaptation action in developing countries (§3)	–	Are remittances a stable enough source of finance allowing for medium to long-term adaptation?
(…) funding as well as **improved access** shall be provided to developing countries (§8)	–	Do remittances provide direct access to funding?
Adequate (…) financial resources (…) to support the implementation of adaptation action in developing countries (§3)	Decision: (…) and **adequate** funding shall be provided to developing country parties (§97)	Could remittances contribute substantially to cover adaptation costs in developing countries?
Adequate funding (…) shall be provided to developing countries (§8)		
Scaled up (…) funding (…) shall be provided to developing countries (§8)	Decision: **scaled-up** (…) funding shall be provided to developing country parties (§97)	Are remittances an increasing flow?
New and additional (…) funding (…) shall be provided to developing countries (§8)	Decision: (…), **new and additional** (…) funding shall be provided to developing country Parties (§97)	Can remittances be recorded as new and additional to former ODA levels?
The collective commitment by developed countries is to provide **new and additional resources** (…) approaching USD 30 billion for the period 2010–2012 (…) (§8)	COP takes note of: (…) developed countries to provide **new and additional** resources (…) approaching USD 30 billion for the period 2010–2012 (§95)	

(continued)

Table 10.1 (continued)

Copenhagen accord	Cancun agreements UNFCCC	Interpretation to analyse remittances
Funding for adaptation will be **prioritized for the most vulnerable developing countries**, such as the least developed countries, small island developing States and Africa (§8)	Decision: (…); funding for adaptation will be **prioritized for the most vulnerable developing countries**, such as the least developed countries, small island developing States and Africa (§95)	Do the most vulnerable developing countries receive relatively large share of remittances?
In the context of meaningful mitigation actions and transparency on implementation, developed countries commit to a goal of **mobilizing** jointly USD 100 billion dollars a year by 2020 to address the needs of developing countries (§8)	COP recognizes: developed country parties commit, in the context of meaningful mitigation actions and transparency on implementation, to a goal of **mobilizing** jointly USD 100 billion per year by 2020 to address the needs of developing countries (§98)	Do developed countries create enabling environments to promote adaptation through remittances?
In the context of meaningful mitigation actions and **transparency on implementation**, developed countries commit to a goal of mobilizing jointly USD 100 billion dollars a year by 2020 to address the needs of developing countries (§8)	COP recognizes: developed country parties commit, in the context of meaningful mitigation actions and **transparency on implementation**, to a goal of mobilizing jointly USD 100 billion per year by 2020 to address the needs of developing countries (§98)	Are remittances a transparent flow? Are remittances transparent from the source to the final users?
The collective commitment by developed countries is to provide (…) resources approaching USD 30 billion for the period 2010–2012 with **balanced allocation between adaptation and mitigation** (§8)	Decision: new and additional resources (…) approaching USD 30 billion for the period 2010–2012, with a **balanced allocation between adaptation and mitigation** (§95)	Do remittances prioritize adaptation over mitigation?

Source: UNFCCC (2009, 2010)

new and additional; adequate; prioritized to the most vulnerable developing countries; mobilized by developed countries; and transparent balanced allocation between adaptation and mitigation),(ii) provides the climate negotiation context explaining how they were distilled from the Copenhagen Accord and the Cancun Agreements and (iii) introduces the angle under which remittances will be dealt to analyze if they can meet the criteria of adaptation finance and be therefore considered in all respects as an 'alternative source'.

Some of these criteria are partly based on longer standing work agreements under the UNFCCC. For example, criteria such as 'new and additional' and 'predictability' have been articulated again and again, not least in Article 4.3 of the

UNFCCC (Müller 2008; Horstmann and Chandani 2011). For climate action–only potentially addressing finance- the Copenhagen Accord includes the additional criteria *'country-driven approach'* and *'based on national circumstances and priorities'* (UNFCCC 2010; §11). Supplementary criteria are proposed by research and climate funds, for example for feasible, effective and efficient adaptation finance (e.g. van Drunen et al. 2009; Müller 2008).

The identified criteria are based on two milestones in UNFCCC negotiations on climate finance: the 2009 Copenhagen Accord and the 2010 Cancun Agreements. The Copenhagen Accord declared to up-scale climate finance for developing countries with USD 30 billion of fast-start finance for the period 2010–2012 and with USD 100 billion per year from 2020 onwards; that the private sector would be one of sources of these financial resources; and started discussions on the Green Climate Fund. However, the Copenhagen Accord itself is a non-binding political declaration: it was brought forward by 114 Parties, but there was no consensus by the Conference of the Parties (COP). One year later, the 196 Parties to the UNFCCC transformed much of the Copenhagen Accords' content on climate finance into COP decision 95–97 of the Cancun Agreements, and therefore these are included in this chapter as well.

Whilst transforming parts of the Copenhagen Accord in the Cancun Agreements, some minor differences were made. For example, the criteria 'sustainable' and 'improved access' are not included in the Cancun Agreements; and 'balanced' only refers to the 30 billion fast start finance period, which ended in 2012. This chapter however still analyses these three criteria, given that they remain important in international climate finance debates. Access modalities and the balanced allocation are for example key concepts in the design of the Green Climate Fund.

10.3 Motivation to Remit and Invest in Adaptation

The International Organization for Migration (IOM) defines remittances as monetary transfers that a migrant makes to the country or area of origin. Most of the time, they are personal cash transfers that can be invested, deposited or donated to a relative or a friend. Although the definition could be broadened further to include in-kind personal transfers and donations (IOM 2009), this chapter focuses on financial remittances only both as private cash transfers and as donations to community projects with a potential to be used for adaptation finance.

Some studies find that remittances are driven by self-interest motives of the sender (Bettin et al. 2012). Others suggest that the altruism motive lead in an increase in remittances to compensate relatives for negative shocks (Agarwal and Horowitz 2002). Starting from these considerations on the motivation to remit, this section discuss the potential for remittances to finance adaptation at community and household level and comply with the 'predictable', 'sustainable', 'improved access' and 'adequate' criteria.

Predictability Although predictable funding is key for developing countries when formulating adaptation strategies and implementing activities (AMCEN 2011; AGF 2010), it is not further defined by neither the UNFCCC, nor in adaptation finance literature. In the Accra Agenda for Action (AAA 2008), predictability is translated into donors strengthening budget planning, thus providing (1) full and timely information on annual expenditure; and (2) regular and timely information to partner countries on their rolling 3- to 5-year forward expenditure and/or implementation plans.

Analyzing this criterion in terms of remittances' potential to comply implies looking beyond traditional donors and focus on private and alternative sources. To this end, 'predictability' is interpreted not as whether the amount of funding decreases or increases, but on whether recipients can anticipate on future adaptation finance, and plan accordingly.

In this context, remittances have proved to be a more reliable source of foreign currency than other capital flows to developing countries such as foreign direct investment and development aid (World Bank 2005). This does not mean that they are not influenced by sudden factors such as economic crises in host countries (Frankel 2011), but their fluctuations to exogenous is quite predictable.

For example, an increase of remittances can be also foreseen in case of economic crises, catastrophic weather events and natural disasters in migrant's origin countries. This shock-absorbing function is emphasized in early literature on the topic corroborating the hypotheses on the use money transfers as risk-spreading and co-insurance mechanisms at family level (Blue 2004). Lately, this practice has been recognized as a strategy to 'help mitigate external vulnerabilities' and 'increase resilience'.

Sustainability This criterion is distinguished from 'predictability' and interpreted as constituted by two aspects: (1) it is replenishes (like a fund) or is self-generating; and (2) it is a stable or increasing flow of financial resources over time. In terms of remittances, the question is whether these are a stable source of finance allowing for medium to long-term adaptation.

In a case study on Morocco, De Haas and Plug (2006) found that bilateral per-capita remittance flows from destination countries only started to stagnate or decline after two decades from the onset of migration. Other studies suggest that migrant remittances tend to reach a peak approximately 15–20 years after migration. With these rates, remittances seem to be a more stable and sustainable source of income than more volatile ones, such as FDI or ODA (with disbursement planning up to 4 years).

Remittances can also be examined for their potential to foster investments with a long-term perspective, which is often crucial in adaptation. Adams et al. (2008) describe how remitters' objectives are divided between the short-term (e.g. food consumption and health needs) and the long-term (e.g. reinforcements of assets and social position). Long-term goals also include income accumulation and increase of economically sustainable livelihood, reduction of exposure to external stresses, food security and more sustainable use of natural resources. As such, remittances have emerged as a key source of livelihood differentiation.

Moreover, these flows are also used to protect people from the destabilizing effects of absent or ill-functioning markets, failing state policies and a lack of state-provided social security (de Haas 2007). For example, an empirical analysis by Giuliano and Ruiz-Arranz (2009) suggests that migrants compensate for the lack of development of local financial markets using remittances to ease liquidity constraints, channel resources toward productive investments and hence promote economic growth in the long-term.

Improved Access should help to use finance more effectively and efficiently. In the context of adaptation, the ultimate goal of improved access is to reach the most vulnerable people. Concrete steps for direct access and enhanced direct access are taken by the Adaptation Fund and the Green Climate Fund (GCF). According to Ayers (2011), vulnerability to the global risk of climate change is locally experienced, which she calls the 'adaptation paradox'. Current governance of funding relationships is often accountable to contributors of climate finance rather than to the most vulnerable people that experience climate change impacts locally (ActionAid 2007). Rather than a discussion on the institutional settings allowing for improved access, under this criterion this chapter thus focuses on whether the most vulnerable and poor have direct access to finance from remittances.

Although mobility has been recognized by the IPCC as a common strategy for climate change adaptation, it is well known that international migration requires a certain amount of resources and remains too costly for the poorest. Those who cannot afford to undertake travels abroad normally engage in internal migration sending remittances likewise to those left behind. The amount, though, is not comparable to international flows, because of the lower wages and currency. However, the distinction between internal and international remittances is very important for adaptation purposes, as those who migrates internally have more opportunities to visit their families and more control on the use of remittances at home as compared to those who have migrated internally.

Evidence exists that these flows are more likely to reach remote areas than private investments motivated by profit-generation. For example, in Ghana and Burkina Faso remittances are used to increase resilience in vulnerable rural areas by supporting adaptation within the farming sector, for instance through the purchase of agricultural inputs (Deshingkar 2011). When 'improved access' is intended as 'easier access', including lack of intermediation, it is more straightforward to examine their impacts. For example, building infrastructure through ODA tend to be several time costlier than it would have been if it was funded by local resources, as foreign aid often requires hiring of international consultants (Acharya 2003). The outcome of the 2015 Finance for Development conference, the Addis Ababa Action Agenda, commits to lowering the transaction costs of remittance flows. If this would be achieved, access to remittances will be even easier.

Adequacy Literature generally interprets 'adequacy' in terms of quantity. For example, Action Aid (2007), Müller (2008), Christiansen et al. (2012) and Flam and

Skjaerseth (2009) refer to sufficiency to cover relevant costs or the inadequacy of adaptation funding compared to the estimated costs. Indeed, van Drunen et al. (2008; 16–17) write that under the Convention, *'adequate (…) funds were meant to help developing countries meet the agreed full incremental costs'*. The question is whether remittances could complement the resources allocated by traditional donors contributing to cover adaptation costs in developing countries.

According to the World Bank, the recorded annual flow of remittances (USD 516 billion) might be a significant underestimate: informal remittances are estimated to be higher in the range of 10–50 % of recorded remittances (Ratha 2003; El-Qorchi et al. 2003). When analysing remittances through their amount, it can be noted how they form a considerable part of the wealth of several countries. For instance, in Mexico remittances are the second largest source of revenues after oil exports (Aparicio and Meseguer 2012). In other countries in different parts of the world, remittances are a vital source of income: they amount to 48 % of Tajikistan's GDP, 25 % of Lesotho's and Nepal's, and 24 % of Moldova's (World Bank 2013).

In certain specific situations, a share of such flows can help to alleviate the impacts of climate change, for example to deal with natural disasters. As shown by the recent evidence in Haiti, it is possible to see that remittances can actually meet the needs for incremental funding better than foreign aid, which seems less sensitive to shocks (David 2010). Remittances seem to have a stabilizing effect in most developing countries vulnerable to environmental changes: by providing a form of private insurance (ex post risk management strategy) and/or by promoting ex ante risk preparedness (ex ante risk management strategy). This hypothesis was tested by Combes and Ebeke (2011) on a large sample of developing countries (113) observed over the period 1980–2007. The results highlight that remittances dampen the marginal destabilizing effect of natural disasters, in particular where remittance ratios comprise 8–17 % of GDP. For remittances, adequacy is not only to be seen in terms of resource quantity, but also for their capacity to effectively flow under particular circumstances, such as climatic risks preparedness and relief.

To summarize: although climate negotiations address adaptation finance at global and national levels and remittances' are not straightforward pledges to adaptation, to some extent they can be considered predictable and sustainable financial flows that can support the most vulnerable people. In fact, under certain circumstances (e.g. shocks or negative trends) literature shows that remittance- flows increased as an effect of the 'altruistic' motivation at the base of certain remit behaviors. This shows how complicated it is to apply criteria ensued by negotiations among states to decisions taken at individual, household and community level.

10.4 An 'Alternative Source' of Adaptation Finance

The ten climate finance criteria are clearly directed towards traditional public finance. In their paper, Pauw et al. (2015) use them to analyse the potential to mobilize private finance for adaptation. In this chapter, remittances are discussed for their peculiarities in comparison to other international streams in view of possibly including them among the 'alternative' sources.

Scaling Up climate finance means constantly increasing it over time, but the UNFCCC does not define by how much and how fast. The increase from the developed countries' USD 30 billion pledge for the period 2010–2012 (i.e. USD ten billion per annum on average) to USD 100 billion per annum from 2020 onwards would be a tenfold increase, or an additional 26 % each and every year up to 2020. Concerning remittances, this chapter analyses to what extent the flows have the potential to be scaled up for adaptation purposes.

While developed countries can only be expected to scale up climate finance if they are confident that these monies will be spent wisely (AGF 2010; 29), diaspora continue to remit regardless. As a matter of fact, the overall annual flow of remittances to developing countries has nearly tripled since 2000 and is also expected to continue at a rate of over 7 % annually from 2012 to 2014 (Kebbeh 2012).

Although remittances grow with around 8 % per year (OECD 2014), this amount cannot be compared with the necessary annual 26 % increase of climate finance. And this potential, cannot be harnessed without the appropriate incentives (e.g. subsidies or tax relief) that make adaptation 'an opportunity', diaspora entrepreneurs will continue focusing on traditional sectors (retail, agriculture, etc.) to invest their extra-money.

'New and Additional' means that climate finance should be new and additional to Official Development Assistance (van Drunen et al. 2008). It can however be discussed whether it should be 'new and additional' to existing, planned or targeted ODA expenditure at the time of the Copenhagen Accord (see Brown et al. 2010). As remittances are not related to a developed-country government budget, it goes without saying that remittances, if used for adaptation purposes, could be recorded as new and additional to former ODA levels. The challenge is to leverage these investments towards adaptation actions and to account for them. Many households might contribute to adaptation without considering it that way (and not knowing that their actions could be supported by further aid devoted for that specific purpose).

Although migrant's financial transfers to their countries and areas of origin are undeniably increasing (World Bank 2014b), it is well-known and acknowledged by most of the international financial institutions that only about 5 % of these flows are used for productive investments. The amount that might be directed towards adaptation actions is thus most likely inferred within this small percentage. We are therefore speaking about a very small part of the huge sum mentioned as remittance flow. Moreover, for this share to be used for future adaptation plans, information is needed, attractive incentives have to in place and depends on the social and cultural context and personal orientations.

The importance of 'alternative' sources is key in the discussions on how to attract new type investors. For this reason, enabling environments for attracting these peculiar investments – done by nationals leaving abroad and targeting adaptation- need to be promoted by governments and their international partners. Remittances might be new and additional sources when the benchmark is the disbursed ODA. However, they cannot be considered as granted, as the direction of their use is very context-specific.

Prioritize the Most Vulnerable Developing Countries Climate funds such as the Global Climate Change Alliance (GCCA), the Pilot Project on Climate Resilience (PPCR) and the Adaptation Fund were all designed to make decisions on country prioritization and allocate funds based on levels of vulnerability, but they all have their own standards for doing so (Klein and Möhner 2011). Altogether it remains unclear what 'prioritization' means in terms of, for example, financial flows or effort made. Of the total public adaptation finance that was approved so far, Climate Finance Update (2014) estimates that 32 % flowed to Africa, 52 % to The least developed countries (LDCs), and 9 % to Small Island Developing States (SIDS); or, given the overlap, 60 % to the three taken together. This hardly reflects a country-based prioritization, considering that these three groups constitute 94 out of 140+ developing countries,[1] and that 22 % of these 94 countries have been excluded from public climate finance interventions so far. A prioritization based on a per capita basis would have very different outcomes, but this chapter analyses along to the UNFCCC outcomes, thus prioritizing on a per-country basis too. This chapter identifies whether the most vulnerable developing countries receive relatively large share of remittances, and installs a 60 % threshold.

The share of all remittances received by today's middle-income countries has risen to an estimated 71 % in 2013 from 57 % in 2000. Although the share to low-income nations has doubled in those years, it remains a small proportion with 6 % of the total (Connor et al. 2013). However, the economic importance of remittances is larger in poorer countries than in richer ones (c.f. Giuliano and Ruiz-Arranz 2009).

Several countries SIDS have important share of GDP constitute by remittances, with the highest amounts in Samoa (23 %) and Haiti (21 %). Among the other most vulnerable groups, Nigeria (Africa) with $21 billion and Bangladesh (LDCs) with $14 billion are among the top recipient countries worldwide (World Bank 2014).

Based on this data, it is impossible to establish a clear-cut connection between the amount of remittances and countries' vulnerability beyond the most vulnerable developing countries as defined by the UNFCCC.

Essentially dealing with the overall amount, the potential share to be invested in adaptation and the countries interested, these criteria go beyond the motivation to remit. Unlike ODA, the quantity of remittances is still growing. Like private investors, remitters respond to incentives to choose specific types of investments (including adaptation) over others (and over consumption). In this context, the role of donors -through e.g. targeted funds, budget support programs and debt swaps- and developing country governments -through e.g. the provision of incentives and fiscal

---

[1] 'Developing countries' is not an official group under the UNFCCC. However, as a comparison: there are 154 non-Annex I parties (see http://unfccc.int/parties_and_observers/parties/non_annex_i/items/2833.php).

easing and the design of legal frameworks- is key to ensure that the right market mechanisms are in place to increase the share of remittances invested in adaptation, as discussed in the next section.

10.5 Channeling Remittances Towards Adaptation: The Role of Governments

In the context of scarce public funds for climate adaptation, the government's role is pivotal in creating an enabling environment for entrepreneurial initiatives and in triggering new resources, including diaspora's investments to build resilience to climate change.

Mobilizing What mobilizing of climate finance entails is neither defined by the UNFCCC, nor in literature. This chapter interprets 'mobilizing' as a pro-active public intervention from developed countries, for example through domestic mobilization of public climate finance, institution building, capacity building, and creating incentives to increase climate financing from other sources. In this chapter, we identify whether developed countries create enabling environments to promote adaptation through remittances.

The increasing amount of remittances and the awareness of the effects that may have on migrants' countries of origin have led both host and home countries to react with a range of public policies. Developing countries with high rates of emigration have already offered incentives to attract and to invest remittances. For example, Senegalese Governmental agencies are promoting diaspora investments in government-run infrastructure projects by offering loans for development projects (Panizzon 2008) and tax exemptions. Since 2008, the NGO FES (La Fondation des émigrés sénégalais) with support by the Ministry of Senegalese Abroad and by Spain, aims at channeling diaspora investments into Senegal (Scheffran et al. 2012). Another example is the Mexican 3 × 1 Program for Migrants, where the: public sector triples the amount of money to encourage the potential investors to choose certain type of projects.

In order for investments to be 'mobilized', however, developed countries have to create a trigger and incentivize such types of investments. They should play an active role beyond employing the migrants. The authors did not find examples in literature. The solution probably lies in developing adequate institutional mechanisms that serve as a basis for cooperation between developed country governments, migrants and potentially international businesses that operate in both the host and the home country.

Transparency Action Aid (2007) suggests that transparency goes beyond purposes (i.e. adaptation), amounts (i.e. USD 100 billion per year), and results of funding (i.e. meaningful), but also includes the governance structure and procedures at providers of financial resources. The Adaptation Fund indeed introduced transparency indicators in its overall management (Horstmann and Chandani 2011). Eventually, transparency on climate finance also means monitoring, reporting, and verification and tracking climate finance from source to final use (Buchner et al. 2011; van Drunen et al. 2009).

As such, transparency is essential to a results orientation and for accountability (Chaum et al. 2011; 2). Just like *'increased transparency in the use of international public finance would elucidate the current and potential role of public finance in leveraging private finance, and would increase understanding of the effectiveness and success rates'* (Brown and Jacobs 2011; 7), transparency on public policies and co-finance aiming to secure or redirect remittances could help to leverage larger spending on adaptation. This will, however, not be easy. An array of unofficial and informal modes of sending money exists (from mailing cash or checks using postal service to the *hawalards*-brokers- scattered across cities, which function as private Remittance Service Provider) and many remain unmonitored (Biller 2007).

In order to harness the potential for remittances towards adaptation finance, the regulatory community requires an approach that meets the goals of financial inclusion and financial transparency. Remittances could increase if legislative barriers and fiscal costs of financial transfers can be reduced; the latter can be facilitated by the introduction of more market players and modes of transmission, better provision of reliable information to migrants on the costs of transfer, and generally better and more credible supervision of the sector (Black 2003). By lacking these conditions, remittances currently do not meet the criterion of transparency. The channels through which they flow are partly informal and not adequately addressed in terms of governance structures and regulations.

'Balanced Allocation Between Adaptation and Mitigation' remains undefined by the COP, but upon their request, the GCF Board decided to *'aim for a 50:50 balance between adaptation and mitigation during the initial phase of the Fund'* (Green Climate Fund 2014; 6). So far, around 16 % of the public climate finance flows to adaptation (Climate Finance Update 2014); the amount of private adaptation finance is very hard to track but seems minimal compared to private mitigation finance (Buchner et al. 2011, 2013). Whether climate finance should be balanced 50:50 between adaptation and mitigation is an open question, but in any case the finance for adaptation needs to increase (see e.g. Terpstra 2013).

Remittances neither principally aim to address climate change, nor do they aim to balance between adaptation and mitigation. However, throughout the chapter we highlighted that remittances can help to increase resilience against climate stresses and that in case of emergencies and disasters, remitters will invest in immediate relief and rehabilitation. Whether this will be translated into adaptation finance and whether diaspora entrepreneurs will invest in long term projects related to adaptation will depend on how each government will set priorities for incentives allocation.

10.6 Conclusion

Although there is extensive literature on the impact of remittances on development, little research exists on their potential to support adaptation to climate change. There is a huge and unexplored potential: recorded remittances to developing countries are expected to increase up to USD 516 billion in 2016 (World Bank2014a; even a small part of which could already be a substantial contribution to adaptation. Furthermore, remittances directly reach the local level, and thus potentially to those most vulnerable to climate change that are difficult to reach through existing channels of ODA and climate finance. And finally, remittances offer opportunities for both climate disaster relief and investments in long-term adaptation.

But rather than looking at whether remittances constitute effective financial means to address adaptation, this chapter addresses the question whether they could also constitute an alternative source of the annual USD 100 billion international climate finance from 2020 onwards, as was pledged by developed countries under the UNFCCC regime. This is not uncontroversial: even if remittances could constitute an alternative source of climate finance, it is ethically questionable whether financial resources of poor migrants can substitute (public) climate finance from developed countries. But in any case, this exercise helps to better understand what alternative climate finance sources could be. Based on empirical evidence from literature, this chapter thus identified to what extent remittances meet ten adaptation finance criteria as negotiated under the UNFCCC Copenhagen Accord and the Cancun Agreement (see Pauw et al. 2015).

This chapter finds that remittances can meet a number of criteria such as 'adequate', 'sustainable', 'predictable' and 'improved access', mostly because they relate to the motivation to invest in countries of origin and, thus, to some extents, to the willingness to protect and support families, friends and communities. It is a matter of personal connection, affection or altruism. Due to these special drivers, remitters are special 'investors' that are available to 'trade off' profit with wellbeing, development and, potentially, adaptation of those left behind in developing countries.

Besides this special feature that remittances might have, these flows remains private flows and, as such, they respond to incentives when considered as stocks of money. Under this lens, criteria such as 'new and additional', 'scaling up' and 'prioritize the most vulnerable developing countries' can be met, but, as any other private source, to be leveraged and channeled towards the aim, there is the need for targeted policies.

Finally, criteria such as 'mobilizing', 'transparency' and 'balanced allocation' are more complicated to be analyzed for the remittance potential to finance adaptation, as they are designed for and typical for public finance. In contrast, remittances are driven by individual interests and market mechanisms and flow regardless to the compliance with these criteria. It is only governments' responsibility to orient them through effective regulations in an attempt for these criteria to be met.

In a first exploration, this chapter found that overall remittances insufficiently meet the ten adaptation finance criteria. Nevertheless, a share of remittances could still meet the criteria and clearly make a contribution not only to adaptation, but perhaps even to international adaptation finance. As a way general forward, the ten criteria in ongoing UN negotiations on climate finance could be altered in order to stimulate alternative sources of climate finance such as remittances. Whether a share of remittances will ever contribute to the mobilization of the annual USD 100 billion of climate finance, and thus constitute 'international climate finance' is, in the end, a controversial political decision.

Acknowledgement The authors express their gratitude to the editors for providing the opportunity to prepare this chapter. The authors would also like to thank the participants and organizers of the COST workshop in Bonn (February 2014) for their comments and suggestions on the concept. Any remaining shortcomings and flaws are solely the responsibility of the authors. Research funding by the German Federal Ministry for Economic Cooperation and Development (BMZ) is gratefully acknowledged

References

AAA – Accra Agenda for Action. (2008). Accra agenda for action. In *3rd high level forum on aid effectiveness, Accra*.
Acharya, S. R. (2003). Official Development Assistance (ODA) in transport sector: Challenges and opportunities. *Eastern Asia Society for Transportation Studies, 4*, 1572–1586.
ActionAid. (2007). *Equitable adaptation finance: The case for an enhanced funding mechanism under the UN framework convention on climate change*. Johannesburg: ActionAid International.
Adams Jr., R. H., Cuecuecha, A., & Page, J. (2008). *The impact of remittances on poverty and inequality in Ghana*. World Bank Policy Research Working Paper Series, vol. (2008).
Agarwal, R., & Horowitz, A. W. (2002). Are international remittances altruism or insurance? Evidence from Guyana using multiple-migrant households. *World Development, 30*, 2033–2044.
AGF. (2010). *Report of the secretary-general's high-level advisory group on climate change financing*. New York: United Nations.
AMCEN. (2011). *Addressing climate change challenges in Africa; a practical guide towards sustainable development*. Nairobi: AMCEN secretariat.
Aparicio, F. J., & Meseguer, C. (2012). Collective remittances and the state: The 3 × 1 program in Mexican municipalities. *World Development, 40*(1), 206–222.
Ayers, J. (2011). Resolving the adaptation paradox. *Global Environment Politics, 11*(1), 62–88.
Bettin, G., Lucchetti, R., & Zazzaro, A. (2012). Financial development and remittances: Microeconometric evidence. *Economics Letters, 115*, 184–186.
Biller, S. (2007). UICIFD Briefing No. 3: Remittances – Iowa City: University of Iowa Center for International.
Black, R. (2003). *Soaring remittances raise new issues*. Migration policy institute. http://www.migrationpolicy.org/article/soaring-remittances-raise-new-issues. Accessed 3 Mar 2016.
Blue, S. A. (2004). State policy, economic crisis, gender, and family ties. Determinants of family remittance to Cuba. *Economic Geography, 80*(1), 63–82.
Brown, J., Bird, N., & Schalatek, L. (2010). *Climate finance additionality: Emerging definitions and their implications* (Policy Brief 2). Heinrich BöllStiftung and ODI.
Brown, J., & Jacobs, M. (2011, April). *Overseas development institute, leveraging private investment: The role of public sector climate finance*. At http:/www.odi.org.uk/resources/download/5701.pdf

Buchner, B., Falconer, A., Hervé-Mignucci, M., Trabacchi, C., & Brinkman, M. (2011). *The landscape of climate finance*. Venice: Climate Policy Initiative.

Buchner, B., Falconer, A., Herve´-Mignucci, M., & Trabacchi, C. (2012). *The landscape of climate finance 2012*. Venice, Italy. Retrieved from http://climatepolicyinitiative.org/wpcontent/uploads/2012/12/The-Landscape-of- ClimateFinance-2012.pdf

Buchner, B., Herve-Mignucci, M., Trabacchi, C., Wilkinson, J., Stadelmann, M., Boyd, R., Mazza, F., Falconer, A., & Micale, V. (2013). *The global landscape of climate finance 2013*. CPI. Venice, Climate Policy Initiative.

Chaum, M., Faris, C., Wagner, G., & Brown, J. (2011). *Improving the effectiveness of climate finance: Key lessons*.

CheSuh-Njwi, A. (2015, August 31). The 'Bush Faller', Akatarian, Bouger': What is the Difference between a migrant and a diaspora? *The African Bulletin*.

Christiansen, L., Ray, A. D., Smith, J. B., & Haites, E. (2012). *Accessing international funding for climate change adaptation*. Roskilde: UNEP Risø Centre on Energy, Climate and Sustainable Development.

Climate Finance Update. (2014). http://www.climatefundsupdate.org/data. Accessed 3 Mar 2016.

Combes, J. L., & Ebeke, C. (2011). Do remittances dampen the effect of natural disasters on output growth volatility in developing countries? *CERDI, Etudes et Documents*, E 2010.31.

Connor, P., Cohn, D., & Gonzalez-Barrerra, A. (2013). *Changing patterns of global migration and remittances* – PEW Research Center 2013. Washington, DC. Retrieved from www.pewresearch.org. Accessed 3 Mar 2016.

David, A. (2010). *How do international financial flows to developing countries respond to natural disasters?* (IMF Working Papers 10/166). International Monetary Fund.

De Haas, H. (2007). *Remittances, migration and social development: A conceptual review of the literature* (Programme on Social Policy and Development, Paper No. 34). Geneva: UNRISD.

De Haas, H., & Plug, R. (2006). Cherishing the goose with the golden eggs: Trends in migrant remittances from Europe to Morocco 1970–2004. *International Migration Review, 40*(3), 603–634.

Deshingkar, P. (2011). Are there examples of remittances being used to build local resilience to environmental change, especially through investment in soil and water conservation or broader agriculture? *UK Foresight*, SR13.

Ebeke, C., & Combes, J.-L. (2013). Do remittances dampen the effect of natural disasters on output growth volatility in developing countries? *Applied Economics, 45*(16), 2241–2254.

El-Qorchi, M., Maimbo, S., & Wilson, J. (2003). *Informal funds transfer systems: An analysis of the informal Hawala system* (Occasional Paper, Vol. 222). Washington, DC: International Monetary Fund.

Flåm, K. H., & Skjærseth, J. B. (2009). Does adequate financing exist for adaptation in developing countries? *Climate Policy, 9*(1), 109–114.

Frankel, J. (2011). Are bilateral remittances countercyclical? *Open Economies Review, 22*(1), 1–16.

Giuliano, P., & Ruiz-Arranz, M. (2009). Remittances, financial development and growth. *Journal of Development Economics, 90*(1), 144–152.

Horstmann, B., & Chardani, A. (2011). The adaptation fund of the Kyoto Protocol: A model for financing adaptation to climate change? *Climate Law, 2*(3), 1–23.

International Organization for Migration. (2009). *IOM and remittances*. http://publications.iom.int/bookstore/free/iom_and_remittances.pdf. Accessed on Feb 2016.

Kato, T., Ellis, J., Pauw, P., & Caruso, R. (2014). Scaling up and replicating effective climate finance interventions. In *Climate Change Expert Group Paper No 2014(1)*. Paris: OECD and IEA.

Kebbeh, O. (2012). *Declining ODA, resilient remittances*. World Bank's blog http://blogs.worldbank.org/peoplemove/declining-oda-resilient-remittances. Accessed on Feb 2016.

Klein, R. (2010). Linking adaptation and development finance: A policy dilemma not addressed in Copenhagen. *Climate and Development, 2*, 203–206.

Lucas, R., & Stark, O. (1985). Motivations to remit: Evidence from Botswana. *Journal of Political Economy, 93*(5), 901–918.

Mohapatra, S., Joseph, G., & Ratha, D. (2012). Remittances and natural disasters: Ex-post response and contribution to ex-ante preparedness. *Environment, Development and Sustainability, 14*(3), 365–387.

Müller, B. (2008). International adaptation finance: The need for an innovative and strategic approach. *Oxford Institute for Energy Studies*.

Organization for Economic Cooperation and Development (OECD) (2014). *Aid to developing countries rebounds in 2013 to reach an all-time high*. Web release 08/04/2014, OECD website.

Panizzon, M. (2008). *Labour mobility: A win-win-win model for trade and development*. The case of Senegal. Geneva Trade and Development Forum.

Pauw, W. P. (2014). Not a panacea: Private-sector engagement in adaptation and adaptation finance in developing countries. *Climate Policy, 15*(5), 583–603.

Pauw, W. P., & Pegels, A. (2013). Private sector engagement in climate change adaptation in the least developed countries: An exploration. *Climate and Development, 5*(4), 257–267.

Pauw, W. P., Klein, R. J. T., Vellinga, P., & Biermann, F. (2015). Private finance for adaptation: Do private realities meet public ambitions? *Climatic Change, 134*(4), 489–503.

Ratha, D. (2003). *Workers' remittances: An important and stable source of external development finance. Global development finance 2003 – striving for stability in development finance Ch. 7* (pp. 157–175). Washington, DC: World Bank.

Scheffran, J., Marmer, E., & Sow, P. (2012). Migration as a contribution to resilience and innovation in climate adaptation: Social networks and co-development in northwest Africa. *Applied Geography, 33*, 119–127.

Straubhaar, T., & Vadean, F. (2006). International migrant remittances and their role in development. In *OECD: International Migration Outlook: SOPEMI 2006*. Paris: OECD.

Terpstra, P. (2013). *Is adaptation short-changed? The imbalance in climate finance commitments*. WRI Blog, 13.11.2013. http://www.wri.org/blog/adaptation-short-changed-imbalance-climate-finance-commitments. Accessed 3 Mar 2016.

The United Nations Framework Convention on Climate Change (UNFCCC). (2009). *Copenhagen Climate Accord*. http://unfccc.int/resource/docs/2009/cop15/eng/l07.pdf. Accessed 3 Mar 2016.

The United Nations Framework Convention on Climate Change (UNFCCC). (2010). *Report of the conference of the parties on its sixteenth session, held in Cancun from 29 November to 10 December 2010*. Bonn: UNFCCC.

United Nations Environment Programme (UNEP). (2014). *The adaptation gap report 2014*. Nairobi: United Nations Environment Programme (UNEP).

van Drunen, M., Bouwer, L., Dellink, R., Gupta, J., Massey, E., & Pauw, P. (2009). Financing adaptation in developing countries: Assessing new mechanisms. In *Climate Change Scientific Assessment and Policy Analysis (NRP-CCWAB)*. Bilthoven: PBL.

World Bank. (2013). *Developing countries to receive over $410 billion in remittances in 2013*. Press release, 2 Oct 2013: http://www.worldbank.org/en/news/press-release/2013/10/02/developing-countries-remittances-2013-world-bank. Accessed 3 Mar 2016.

World Bank. (2014a). *Remittances to developing countries to stay robust this year, despite increased deportations of migrant workers*. Press release, 11 April 2014 http://www.worldbank.org/en/news/press-release/2014/04/11/remittances-developing-countries-deportations-migrant-workers-wb. Accessed 3 Mar 2016.

World Bank. (2014b). *Migration and remittances: Recent development and outlook* (Migration and Development Brief 22).

World Bank. (2005). *Global development finance 2005*. Washington, DC: World Bank.

Yang, D. (2008). Coping with disaster: The impact of hurricanes on international financial flows, 1970–2002. *The B.E. Journal of Economic Analysis & Policy, 8*(1 (Advances)), Article 13.

Chapter 11
Conclusion: Migration as Adaptation: Conceptual Origins, Recent Developments, and Future Directions

Robert McLeman

11.1 Introduction

> *What we do know is that mobility and migration are key responses to environmental and non- environmental transformations and pressures. They should therefore be a central element of strategies of adaptation to climate change. This requires a radical change in policy makers' perceptions of migration as a problem and a better understanding of the role of local and national institutions in supporting and accommodating mobility.* (Tacoli 2009, abstract)

This statement by Cecilia Tacoli is one of the wisest things I have read on the subject of environment and migration, and sets the tone for what follows.

A decade ago my colleague Barry Smit and I advanced a similar argument, though not as eloquently as Tacoli, in an article in *Climatic Change* with the title "Migration as an adaptation to climate change" (McLeman and Smit 2006). At the time we did not expect that our article would be cited by other scholars as often as it has been, or that this particular way of conceptualizing the relationship between human migration and the environment would become so widely accepted. The editors of the present book have kindly invited me to reflect on how this conceptualization has evolved over the subsequent decade, and to consider the future directions research and practice may take. I am also going to offer some comments on the origins of the concept so far as I understand them. However, before going any further, I should emphasize that although I may have been among the first to explicitly describe 'migration as adaptation' in published work – beginning with a public commentary written for the Canadian Security Intelligence Service (McLeman and Smit 2003) – other researchers were reaching similar conclusions around the same

R. McLeman (✉)
Department of Geography & Environmental Studies, Wilfrid Laurier University, Waterloo, ON, Canada
e-mail: rmcleman@wlu.ca

time. Had Professor Smit and I not chosen such a title for our work, someone else would have inevitably done so. If migration as adaptation – or MAA, for short – is starting to become a paradigm, we take no credit for establishing it; the paradigm was already emerging when we wrote of it.

The notion that the decision to migrate could or should be described in terms of adaptation is a relatively new idea in social science research on migration, but is only a small variation on theories of behaviour that have long been embraced by researchers in the natural sciences. The simplest explanation for why MAA has become so widely accepted is because (1) there is growing concern that anthropogenic climate change will force millions of people worldwide to migrate and (2) the main policy instrument for dealing with such possibilities – the UN Framework Convention on Climate Change (UNFCCC) – explicitly requires us to frame discussions of the human impacts of climate change (including migration) in terms of the adaptation of vulnerable people. Had the UNFCCC instead used terms like, say, resilience or capability, the present chapter might instead be a reflection on how 'migration as resilience' or 'migration as capability' became such a widespread term. For those who wish to read no further, that is the short version of how we came to use this terminology so widely; like any good academic, I will now offer a more detailed explanation.

11.2 Early Origins of the Term

In the context of migration causality, the term *adaptation* has its origins in the natural sciences, and more specifically in the theory of natural selection proposed by Charles Darwin (1859). In Darwinian Theory, adaptation refers to physiological or behavioral changes in organisms that increase the likelihood of individuals successfully reproducing and transmitting their genes to the next generation, which in turn cumulatively increases the likelihood of the species as a whole expanding its range, maintaining or growing its population numbers, and otherwise competing successfully with other organisms. It is understood in the natural sciences that there is no 'perfect' adaptation, since the environment to which an individual or species is adapting is continually changing, as are the adaptations of its competitors. Migratory behavior among non-human species is therefore seen by natural scientists as a form of adaptation. Indeed, a simple search in a scholarly database using the terms 'migration+adaptation' without the modifier of '+human' will return studies not only about people, but also trees and salmon and parasites and any number of other non-human organisms.

Unlike humans, migratory behavior by other organisms entails a combination of behavioral and physiological adaptations. This is because when other organisms migrate, they obtain the information needed to guide their movements not from agents or Google maps or word of mouth from social contacts, but by interpreting signals in the natural environment. Many animals possess an 'internal compass' that tells them where and when to travel, based on cues such as variation in the Earth's magnetic field, the polarity of sunlight, gravitational pull, and scents, among others (Able 2005). Although there have been instances when scholars have accused particular groups of people as being inherently migratory in disposition or nature,[1] there is of yet little scientific evidence to suggest that migratory behavior among humans is reflective of any physiological or genetic inheritance. Genomic research may one day prove otherwise, but at present, if we are to apply natural science principles to human migration, it is safest to assume that it is a behavioral adaptation shaped by socio-economic and cultural processes, and not genetic ones.

Research on human migration in the western social science tradition stretches back to Ravenstein's inquiries of the causes and effects of migration, published in the late 1880s in various incarnations under the title 'Laws of Migration' (Ravenstein 1889). Ravenstein would have been well aware of Darwinian Theory, but he did not see migration as being related to adaptation, nor did other migration scholars for most of the following hundred years. Where researchers did use the term adaptation, it was typically done in reference to the process by which migrants adapt physically, socially, and even biologically to living in new communities, surroundings, and cultures (Price 1968; Little and Baker 1988). The term adaptation was almost never used in explicitly in the context of migration causality, although it was implicit or implied in aspects of once-popular theories such as social Darwinism (Spencer 1860), the state's need for *lebensraum* (Ratzel 1902), and environmental determinism (Semple 1911; Huntington 1924). By the end of the Second World War, these overtly Darwinistic social theories had become largely discredited, and the number of published investigations of the role of the natural environment in migration causality declined for several decades. An occasional study of how migrants might have preferences for particular environmental conditions at their destination would appear now and again, but very few explored how environmental conditions might stimulate people to migrate out of particular places (see Svart 1976 for a review).

Environmental influences on migration causality began slowly reappearing in migration scholarship in the latter half of the twentieth-century, with Hunter (2005) tracing this reappearance to Petersen's (1958) general typology of migration, in which he suggests migration can in some circumstances be an *innovative response* to ecological 'push' factors. Wolpert (1966) subsequently identified migration as being an *adjustment* to environmental stress. These terms of 'innovative response' and 'adjustment' foreshadow the idea of MAA, but it would still be two more decades before either environmental push factors or MAA would spark any great amount of interest among scholars. In the intervening period, when environmental factors were considered in the context of migration, it was still most often to look at how environmental conditions and amenities act as 'pull' factors that attract migrants, such as the attraction of sunny climes for internal migration in the US and Australia (Watkins 1978; Biggar 1979; Mullins 1979).

---

[1] For example, some scholars have suggested that rural Oklahomans of the 1930s were an inherently transient, migratory people, although this is inconsistent with their behavior once they reached California, where most settled down permanently as soon as socio-economic circumstances permitted them to do so. See Manes (1982) and McDean (1986).

11.3 The Emergence of MAA

In the late 1970s and early 1980s, the field of natural hazards research, which had been dominated by physical scientists employing primarily quantitative research methods, experienced a social science turn. Researchers like Burton et al. (1978), Blaikie (1981), and Hewitt (1983) began questioning the steadily-mounting property losses, human suffering, and casualties caused by environmental hazards, environmental degradation and soil erosion, especially in low-income countries, despite the tremendous amount of scientific knowledge about the causes and the appropriate methods for reducing loss and harm. Augmented by Sen's entitlement theory research on the origins of famines (Sen 1977), the resulting literature on the political ecology of natural hazards and environmental degradation realized that the potential for loss or harm was not simply a question of exposure to physical risks, but also the relative vulnerability of those exposed and their capacity to adapt (for a more detailed review, see Adger 2006). Large-scale displacements of people in Asia and Africa in the wake of extreme storms and severe droughts in the 1970s and 1980s provided visual evidence for political ecologists that the challenge was not a lack of knowledge, but a lack of means. In this context, migration and relocation were seen not as adaptations but as evidence of a lack of adaptive capacity.

This view of migration as reflecting an inability to adapt was further reinforced by the concurrent 'environmental refugee' literature that was emerging in the NGO and multilateral sectors (El-Hinnawi 1985), which argued that environmental events like extreme weather, human-induced environmental degradation like soil erosion, conflicts over natural resources, and the construction of large dams were becoming important drivers of forced migration in low income countries. This in turned spawned a new sub-discipline of environmental security research, which re-examined cases of political instability, violence, and forced migration, and concluded that environmental factors like resource scarcity played a causal role (Brown et al. 2007).

In the midst of this burst of interdisciplinary research on the relationship between human-being and environmental stressors emerged conclusive evidence of climate change as a new threat to the global environment. As concern grew that anthropogenic greenhouse gas (GHG) emissions are stimulating changes in the Earth's climate system, the UN General Assembly tasked the World Meteorological Organizations and the UN Environment Programme with creating an Intergovernmental Panel on Climate Change (IPCC) to advise global policymakers on developments in scientific research and options for controlling GHG emissions. The IPCC self-organized into three Working Groups, the first to focus on the physical science of how the atmosphere responds to anthropogenic forcing, the second to look at the impacts of climate change, and the third to look at ways of mitigating GHG emissions. In its early years, the IPCC was dominated by natural and physical scientists, and in the first IPCC Assessment Report in 1990, Working Group II – responsible for explaining the impacts of climate change – explicitly imported from natural hazards research the terms *vulnerability* and *adaptation* to describe the implications of climate change for human settlements. References to migration in that report consisted of a few general warnings about environmental refugees and a growing risk of displacements from natural hazards and sea level rise, but with little scholarly evidence to substantiate them. This report provided the backdrop against which the UNFCCC was negotiated in 1992, when signatories explicitly pledged in Article 4.1.e to "cooperate in preparing for adaptation to the impacts of climate change". With this single clause, any actions taken in response to the impacts of climate change – including the relocation, resettlement, or migration of people – would subsequently become in UNFCCC parlance "adaptation".

Throughout the 1990s and early 2000s, the IPCC reporting process placed a growing emphasis on describing what successful adaptation to climate change means and how best to achieve it. By the time of the Third Assessment Report (TAR) in 2001, migration had come to receive its own fifth-order sub-headings in three different chapters and, while continuing to reiterate warnings about mass displacements and environmental security, the IPCC began modifying its perspective on the role of migration. Specifically, the IPCC notes in TAR at Chapter 11, subsection 11.2.6.1:

> Migration in itself is not necessarily a signal of vulnerability to present-day extreme events. Motivations for migration are diverse...

and, later that same section

> Climate change will act in parallel with a complex array of social, cultural, and economic motivations for and impacts of migration.

Here we see the IPCC moving away from the environmental refugee/push-pull approach to describing environmental migration, and toward one that considers migration in ways more keeping with the wider social science approaches to understanding migration, albeit with the added dimension of environmental change thrown into the mix. Migration was still being be treated as an adaptation, but now with recognition that it is an adaptation strongly influenced by social processes.

11.4 Migration as Adaptation as Initially Conceived

It was against this backdrop that my own research on MAA began to develop. As mentioned earlier, I was not the only one working on the topic, but since I cannot speak for others how they came to the topic, I will offer a brief explanation of how I did. In 2002 I began doing doctoral research on environmental migration with Professor Smit at the University of Guelph, after having served abroad for more than a decade as a Canadian Foreign Service officer, posted to places like the Balkans, India, and Hong Kong, where managing migration to Canada formed much of my daily work. My plans for my thesis research had been to travel to Central America and study the long-term migration outcomes of Hondurans displaced by Hurricane Mitch in 1999. In a graduate course I did a research paper on the Dust Bowl migration of the 1930s, when hundreds of thousands of people left the drought-stricken US Great Plains for the Pacific Coast. It dawned on me that, although the event took place decades earlier, there were still people alive who had experienced that particular migration event firsthand, and who might be able to shed some light on the distinction between those who had left the drought stricken areas and those who had remained behind. Were I to have delayed even a few years, most of those remaining eyewitnesses would have passed on.

Although the droughts of the Dust Bowl years cannot be attributed to anthropogenic climate change, it had been argued convincingly by other scholars that the physical conditions experienced during the 1930s were analogous to the sorts of conditions to be expected under climate change for that region and for other dryland regions around the world in coming decades (Glantz 1991; Rosenzweig and Hillel 1993). It was therefore reasonable to assume that, although subsequent socio-economic and technological changes would need to be accounted for, this case could provide insights into the process of migration decision-making under conditions of environmental stress, and identify factors that distinguished those who migrate from those who do not.

I spent a considerable amount of time in declining small towns in Oklahoma and dusty agricultural settlements in California's San Joaquin Valley, reviewing historical records and interviewing old-timers who had experienced what they called the "Dirty Thirties". It became clear to me that the environmental refugee paradigm and its simple push-pull explanations did not accurately reflect what happened in the 1930s. Despite enduring the worst combination of environmental and economic hardship in living memory, residents of the Dust Bowl region exercised considerable agency in their responses to the impacts. While many migrated away from the drought stricken regions, many more did not. Most people tended to resist migrating away from the region until other possible forms of adjusting and coping had been exhausted; even then, many refused to leave. Some who did migrate left temporarily while others left for good. Still others migrated into the drought-affected areas because they were fleeing urban poverty, and thought they might at least grow their own food. Most significantly, those who left for good were not a random representation of the population of the drought stricken area; rather, they possessed particular cultural, social, and economic capital that made migration a viable response.

In short, though the Dust Bowl migration has often been characterized as an 'exodus' of involuntary migration of the environmental refugee type, it was more an outcome of conscious decision-making by people who assessed in 'real-time' the options available to them, and made the choices that best suited them under trying conditions. In other words, it was a process of adjustment and adaptation. This does not mean to say that there were no incidents of people having to live under refugee-like conditions. There were indeed people in Oklahoma who suffered from malnutrition, and who lived in shantytowns on the edge of garbage dumps or under bridges in self built homes because they had no place else to go. Typically, these were the poorest of the poor, often infirm, elderly, or unwell; women and children who had lost or been abandoned by their spouses; and others who slipped through the cracks of a society that had yet to develop or implement any of the basic social safety nets

Americans today take for granted. They might have wanted to migrate away from the drought-affected region, but lacked the means to do so.

Once I finished reporting my empirical findings from this study (McLeman 2006), Professor Smit and I began to consider how these might inform an understanding of the relationship between climate change and migration more generally. We were well aware of the UNFCCC lexicon and its emphasis on vulnerability and adaptation, and concluded that migration could indeed be situated comfortably within that terminology. Why? In simple terms, the things that the IPCC and climate change researchers saw as being important determinants of adaptation and adaptive capacity – access to financial resources, educational attainment, social networks, institutional arrangements, and so forth – were essentially the same things that migration researchers had long seen as being important influences on migration behavior, and which I had observed in my empirical work. In other words, migration and adaptation were consistent in conceptual terms and, in at least this one empirical case study, in reality as well. Our subsequent article in *Climatic Change* on the subject was timely, as it was referenced in the IPCC's 2007 assessment report in a special feature box dedicated to migration as adaptation.

11.5 The Expansion of MAA in Research

The IPCC's 2007 assessment report provided an initial catalyst for the growth in MAA-situated research in the years that followed. There were studies that considered the concept itself and discussed whether to treat it as MAA or migration as failed adaptation (e.g. Perch-Nielsen et al. 2008; Tacoli 2009; Bardsley and Hugo 2010), as well as empirical investigations that showed MAA being practiced in locations as disparate as the African Sahel and atolls of Tuvalu. In the Sahelian example, researchers (Barbier et al. 2009) noted that vulnerability and adaptive capacity are not simply functions of poverty; for example, poor farmers in northern Burkina Faso are not necessarily always vulnerable to droughts, despite it being a common phenomenon there. The authors also found that rural populations were not as vulnerable as outsiders perceived, but were engaged in a continuous process of adapting to changing environmental and socio-economic realities. Adaptation strategies promoted by outside governments and NGO agencies were rarely as successful or acceptable to residents as were strategies residents developed on their own. Permanent and temporary migration to internal and international destinations formed part of the broader suite of adaptation strategies employed by rural households. While there is always a steady flow of migrants in and out of the region, the authors observed that after a severe drought in 2004 there was a surge in the number of households sending temporary migrants elsewhere, followed by increasing returns in remittance income. A different study of migration in the western Sahel by Scheffran et al. (2012) came to similar conclusions, finding that social networks, migration, and remittances are key factors in creating adaptive capacity and fostering local economic development.

In the case of Tuvalu – a nation of small atolls widely suspected of already being affected by the impacts of climate change – researchers (Mortreux and Barnett 2009) found that the high numbers of people migrating abroad were doing so to seek employment and/or join relatives in more developed countries. Although most study participants were aware of the potential impacts of climate change, they intend to adapt to it in ways other than migration as much as possible. In a situation such as theirs, the question of whether any eventual migration should be interpreted as an adaptation or a failure of other forms of adaptation is very much an academic one. It is also worth noting that another atoll nation, Kiribati, has been advocating for a 'migration with dignity' strategy that would see its young people have greater ability to migrate to developed nations, allowing them to develop skills, accumulate wealth, and remit funds that enhance the nation's capacity to adapt and, should relocation become essential, would give them greater agency over their migration so that they do not become 'environmental refugees' (Kiribati Office of the President 2015).

Another important catalyst for a MAA-orientation in academic research emerged with the final report in 2011 of the Foresight Project on Global Environmental Migration, sponsored by the UK Government Office for Science. This multi-year research initiative involved hundreds of contributors in over 30 countries, and sought to understand how environmental factors might affect the volume and patterns of human migration in coming decades, and identify strategies that might be implemented now to prepare for and manage such eventualities. In their executive summary of the study's outcomes, the authors write at page 10:

> Migration can represent a 'transformational' adaptation to environmental change, and in many cases will be an extremely effective way to build long-term resilience. International policy should aim to ensure that migration occurs in a way which maximises benefits to the individual, and both source and destination communities. (Foresight 2011)

The title of a follow-up commentary submitted to *Nature* by the Foresight project's lead authors summed up their conclusions succinctly: Migration as Adaptation (Black et al. 2011). The Foresight report and its MAA approach has since become an important touchstone for environmental migration researchers, and helped spark interest in the subject across a wide range of disciplines, which importantly now include law, policy studies, and related fields.

11.6 The Formalization of MAA in International Climate Policy

At the same time that academic research on MAA was expanding, adaptation was taking on greater importance within the UNFCCC process. Starting in 2001, funding had been made available through the Global Environment Facility for least-developed countries[2] to implement initiatives in response to short-term impacts and risks associated with climate change via National Adaptation Programmes of Action (NAPAs). In 2010, signatories to the UNFCCC decided at their meeting in Cancun that adaptation should be given equal priority to mitigating GHG emissions, and encouraged all signatory nations, not just least-developed countries, to formulate long-term National Adaptation Plans. In addition, in what became known as the Cancun Adaptation Framework, the Conference of the Parties to the UNFCCC wrote in Section II, article 14 that it:

> 14. Invites all Parties to enhance action on adaptation under the Cancun Adaptation Framework, taking into account their common but differentiated responsibilities and respective capabilities, and specific national and regional development priorities, objectives and circumstances, by undertaking, inter alia, the following:
>
> ...
>
> (f) Measures to enhance understanding, coordination and cooperation with regard to climate change induced displacement, migration and planned relocation, where appropriate, at the national, regional and international levels;

With this, the UNFCCC signatories sent a clear signal that going forward migration should formally be considered a form of adaptation to climate change, and should therefore form part of countries' long-term adaptation planning where appropriate. It also had the effect of recasting environmental migration in a somewhat less negative light. Before Cancun, least-developed countries in their NAPAs tended to represent migration (especially internal rural-urban migration) as being problematic, something to be controlled and avoided (Warner et al. 2015). The new longer-term National Adaptation Plan approach provides an opportunity to build migration and mobility more systematically into formal adaptation plans, to obtain funding to support adaptive migration, and to seek favourable outcomes from MAA. Researchers and policymakers have since been actively engaged in envisaging what sorts of policy and legal instruments might be developed to make such things possible. Some better-known examples include the Peninsula Principles on Internal Displacement and the Nansen Initiative on Cross-Border Climate-Induced Displacement. As there are much better analyses of these already available (e.g. Gemenne and Brücker 2015), I will skip on ahead to some thoughts on the future of MAA research.

11.7 Future Opportunities in MAA Research

11.7.1 Theoretical and Conceptual Development

There are many directions research on MAA might take in coming years; a few important priorities spring immediately to mind. A first concerns the theoretical and conceptual development of MAA. Despite its relentless march toward becoming the

---

[2] The term "least-develop country" is one that is explicitly used in the UNFCCC. For a list, see http://unfccc.int/cooperation_and_support/ldc/items/3097.php

dominant paradigm in research and policymaking, MAA is not without its critics. This is a good thing. All paradigms need to be continuously challenged and critically appraised, and MAA is no different. As an example of recent critique, Felli and Castree (2012), writing after the release of the Foresight report, suggested that MAA simply reinforces the larger neo-liberal project of incorporating ever larger numbers of people into capitalist labour markets where the poor are inherently disadvantaged. Further, the authors find the emphasis on households, communities, and local social networks as being the primary agents of adaptation problematic, because it treats state-level politics as being almost exogenous to the adaptation process, letting states off the hook for responsibilities they ought to be assuming. Bettini (2014) offers a similar critique, suggesting that although many scholars (myself included) champion the move away from the language of environmental refugee and environmental security to the MAA paradigm as being a positive development, the result is a further strengthening of neoliberal valuation of migration and adaptation. Another set of critiques draws attention to the use of language implicated in MAA. For example, Baldwin (2013) has pointed out that the terms and vocabulary we use to describe environmental migrants can have embedded racial connotations, which could do with having more sunlight shone upon them. Mayer (2015) questions whether we wish to make the climate migrant a privileged class in research and policy, more deserving of protection and adaptation assistance than other people.

Such critiques are valid, although complaints about MAA feeding the neo-liberal machine might equally be levelled at just about any form of human migration undertaken today (environmentally related or otherwise), apart from that pursued by the few traditional societies that still remain. MAA is heavily implicated in (and influenced by) policymaking; as policymakers tend to be neoliberal in outlook, it is perhaps inevitable that MAA-based policy and research that feeds into it will have a neo-liberal orientation. I, too, have observed that there is a distinct "us-and-them" undercurrent in discussions about environmental migration, a reason why I have made a point in my own research and writing to remind North Americans that environmental displacement and migration can happen and has happened to us, here on our continent, too. I also agree that care must be taken in adaptation planning to consider those who cannot or do not need to migrate as much as we do those who do migrate, an observation made elsewhere by Black et al. (2012), among others.

The more challenging critique, in my own view, is that of the role of the state. A recurrent theme in MAA research is that undesirable[3] environmental migration tends to happen most often in locations and jurisdictions where the state and its agents are unprepared, incompetent, corrupt, or worse. While such conditions are common in many least-developed countries, they are not exclusively so. The disastrous response to Hurricane Katrina shows that least-developed states have no

---

[3] I use the term 'undesirable' deliberately, for not all environmental migration is unwanted or unbeneficial. Retirees moving to sunnier climes of young people from Australia and Japan moving to Whistler for the skiing are examples of desirable environmental migration; desirable from the perspective of the receiving area and, most importantly, of the migrants themselves.

monopoly on incompetence when it comes to managing environmental hazards and post-event migration (see McLeman 2014 for a more detailed analysis). A reality is that, if states were to act consistently and exclusively in the interests of their citizens, environmental migration would occur only in those situations where there is little other choice, such as the abandonment of the capital city of Montserrat in 1997 due to volcanic eruptions. A challenge for theoreticians and empirical scholars alike is to find ways of pushing states toward taking greater responsibility and accountability for building adaptable societies, and creating the socio-economic conditions under which MAA is less necessary and is undertaken with the highest degree of possible agency when it is necessary.

That said, no amount of competency and benevolence at the level of the state will entirely prevent undesirable environmental migration in the future. For example, given the inability and unwillingness of the global community to make meaningful reductions in GHG emissions, it is a question of when, not if, residents of atolls in Kiribati will need to be relocated because of rising sea levels, coastal erosion, and groundwater salinization. Adaptation through migration will at that point become the only option, full stop. The pursuit of a 'migration with dignity' strategy by the people of Kiribati, and the support of scholars for such a policy (e.g. Wyett 2014) thus makes good sense. At the same time, such predicaments open up questions of what citizenship and human rights mean in a future where sovereign territories cease to be viable and their residents become subject to the mercies of an international community to whom human rights seem increasingly subject to negotiation. Maldonado et al. (2013) have already begun raising such questions in the context of indigenous communities being displaced by coastal erosion in Alaska; more research in other regions seems warranted.

11.7.2 Empirical and Methodological Challenges

There are many opportunities for improving our empirical knowledge of environmental migration in general and MAA in particular, and room for further innovation in our methodological approaches. The number of empirical investigations has been growing rapidly in recent years, but despite this, the overall body of evidence remains quite limited, especially on the subject of international migration for environmental reasons (Obokata et al. 2014). In recent years I have been working with colleagues to understand and document the influences of environmental events and conditions abroad on international migration to Canada. Unlike other nations, migration to Canada comes primarily from non-contiguous countries in Asia, the Middle East, the Caribbean, and Africa. What we are finding is that, while environmental factors abroad do have an influence, that influence is highly specific to particular countries and regions of origin, and is usually a second- or third-order causal factor after labour market or family reunification considerations (Veronis and McLeman 2014; Mezdour et al. 2015). Unlike the environmental refugee hypothesis would suggest, environmental hazards and degradation in places like the

Philippines, Haiti, and Bangladesh are not driving migrants to Canada; rather, they provide additional impetus for those who might already be contemplating migration. This does not mean there is no environmental migration in those countries; there is, only those most affected are migrating internally, and lack the means to reach expensive, far off destinations like Canada. It also does not mean there are no people who move to Canada primarily for environmental reasons; there are, but their numbers are small enough that they arrive unobserved within the larger flow of labour migration to Canada.

This last point – the unobserved environmental migrant – warrants additional attention. I suspect there are many people on the move today for whom environmental degradation, natural disasters, and/or pollution provide a key impetus for their decision to migrate. They are adapting to environmental stress via migration, but we do not notice them because when asked by officialdom to give their reasons for migrating, (a) officialdom does not ask about environmental motivations and/or (b) there is no advantageous reason for the migrant to declare environmental reasons, when the ability to stay in the destination area depends on being classified as something else (or avoiding officialdom entirely, and becoming an undocumented migrant). And it is not only the international environmental migrant who goes undetected. Ackerly (2015) provides a compelling argument that in places like Bangladesh, the most vulnerable people, those 'trapped' populations who lack the means to flee environmental degradation, or those who do but remain mired in poverty, are often 'hiding' in plain sight. We become so accustomed to seeing poor rural women walking long distances to collect water and fuelwood, or families living in temporary shelters awaiting floodwaters to recede that we fail to even notice them or inquire as to what brought them to such a position.

This in turn raises another issue warranting further empirical inquiry: the relationship between socio-economic inequality and MAA. Each contributes to the other in a dialectic way (McLeman et al. 2015). There seems to be general agreement in the literature that in situations where environmental migration might occur, there are three categories of people: those who have the means to migrate but elect not to, because they have other means of adapting; those who might want to migrate but lack the means to do so; and those who have the means to migrate and do so (i.e. those who are the MAA adapters). But how well do we know the combinations of environmental and socio-economic conditions that determine membership in these groups? Not very well. And what are the impacts of MAA out-migration on the people and places left behind? Again, more information is needed.

Another important empirical question is that of thresholds. Bardsley and Hugo (2010) wondered if there are critical thresholds we can identify that, once crossed, make migration a vital component of adaptation when previously it was not. To answer this question and the ones in the preceding paragraph, we have to first determine if it is possible (and wise) to make conclusions by comparing across disparate case studies or 'scaling up' from local case studies to make regional or global conclusions. We are not yet at the point where we have a robust enough body of empirical evidence to do so reliably, but we are getting there.

Assuming we do get there, the next question policymakers will be demanding of researchers is, can we quantify MAA and attribute a value to it? If we can do so, that is, if we can identify under what conditions MAA is 'good' and when it is 'bad', can we then make prescriptions of how to increase the preconditions that make MAA good, and reduce those that make it bad? Researchers will likely also be asked to evaluate the desirability and social acceptability of MAA relative to other forms of adaptation; after all, if the international community is going to be paying out large amounts of money in adaptation assistance under the UNFCCC, it will want to know where and how to get the most value for its money.

11.7.3 MAA and Policy

A third area where there will be considerable growth in research in coming years is that to be done in support of policymaking. In several places above I have mentioned there is growing interest in finding ways to shrink down the number of people who will be obliged to migrate for environmental reasons in the future, and to maximize the agency of those who do. While all forms of environmental hazards and degradation have the potential to create unwanted displacement and migration, sudden-onset events like tropical cyclones merit particular attention. Research and innovation in monitoring, early-warning systems, and evacuation planning have successfully brought down the death tolls from sudden-onset disasters in recent decades. However, the combination of unexpected onset and lasting post-event damage to housing stocks continues to make such events particularly capable of generating distress migration, and of undoing years of economic development, as hurricanes Mitch and Katrina showed. If such events are to become more frequent and/or destructive as is expected under climate change, more policy-related research on the relationship between housing and post-disaster displacement and migration is warranted. Identification of low-hanging fruit in terms of building adaptive capacity proactively would be especially useful.

In the case of slower-onset events like drought and land degradation, there seems to be a lethargy in the responses of governments and the international community, almost as if there is a desire to wait and see how bad things get before states and institutions (and even individuals) take action. This certainly seems to be the case in western North America, where despite an abundance of scientific evidence about the severity of the current drought and the precariousness of western water supplies, governments continue to make only the smallest of adjustments and adaptations. It is a similar case when least-developed countries are stricken by drought and turn to the international community for help; the response they receive is typically underwhelming at best. This is a case where the research community is warranted in being more normative, if not hectoring, in its tone with policymakers. International policymakers must get away from ad hoc/wait-and-see approaches to dealing with droughts and long term changes in precipitation; the evidence for this has been made abundantly clear in the findings of the recent "Where the Rain Falls" research project on precipitation-related migration (Warner and Afifi 2014). The next step to be encouraged therefore is more research on how to translate this good science into good policymaking.

More research is also warranted on how to create and implement labour migration agreements that support the 'migration with dignity' agenda described earlier. Fostering adaptation through labour migration seems on face value to be a win-win proposition, since in most developed countries labour forces and populations are aging and shrinking, while in highly vulnerable, less developed nations populations are typically young and increasing fast. The opportunity here is for researchers to provide evidence that such initiatives would be beneficial, and identify the arrangements under which they would work best.

Some research is already being done to provide evidence and guidance to policymakers on the governance structures that would make MAA successful. For example, Geddes and Jordan (2012) analyzed European policies and programs, and found them to be poorly coordinated and unhelpfully security-oriented. The authors recommended greater cooperation within the EU, the development of more partnerships with countries outside the EU, and for policymakers to see migration as a positive form of adaptation and not as a problem. Although they were looking ahead to future environmental migration, the authors' diagnosis and prescriptions seem prescient given what has transpired in 2015 with the influx to Europe of migrants from Syria and elsewhere.

A final related question is how best to incorporate MAA within the UNFCCC process. Various suggestions have already been made, from suggestions of separate, formal accords on environmental migration (Biermann and Boas 2012) to mainstreaming MAA within other adaptation arrangements under the UNFCCC itself (Gibb and Ford 2012). There remains considerable room for additional research, discussion and debate on this subject.

11.8 Conclusion

There are many more needs, avenues, and opportunities for future research on MAA that have not been mentioned above; what has been offered should be seen as a sampling intended to whet the reader's appetite, and not a complete menu. My representation of the evolution of MAA has likely omitted many important twist, turns, and developments, and has named far too few of the many scholars who have contributed to its development. I have also no doubt omitted many important criticisms and critiques of MAA that are worthy of discussion. I hope the reader will forgive these oversights; although I have often spoken and written in favour of MAA, I do not pretend to know all of its intricacies.

What I believe I have demonstrated with reasonable coherence is that, like it or not, MAA is here to stay. It is strongly embedded in adaptation processes and mechanisms under the UNFCCC, and there are no signs it is likely to be replaced by anything else in the near future. Environmental migration research has co-evolved with international climate change policymaking, and has been generally accepting of MAA as a departure point for theoretical and empirical inquiry. This co-evolution makes environmental migration research somewhat distinctive as compared with other types of natural or social science research. In many other specialties, researchers bemoan the lack of attention paid to their work by policymakers. That is not the case here. Environmental migration researchers have the ears of policymakers if they wish to make themselves heard. This does, however, come with the caveat that the message must be communicated in the language of vulnerability and adaptation that policymakers have come to expect. Embracing MAA thus simultaneously creates opportunities and imposes constraints on the researcher. So long as we are aware of these opportunities and constraints going forward, there is considerable potential for academic and policy growth and development in this field.

Acknowledgement The author's research on environmental migration is actively supported by an Insight Grant from the Social Science and Humanities Research Council of Canada.

References

Able, K. P. (2005). Migratory behavior. In E. Geller (Ed.), *Concise encyclopedia of environmental science* (pp. 426–427). New York: McGraw-Hill.

Ackerly, B. A. (2015). Hidden in plain sight: Social inequalities in the context of environmental change. In R. McLeman, J. Schade, & T. Faist (Eds.), *Environmental migration and social inequality* (pp. 131–150). Dordrecht: Springer.

Adger, W. N. (2006). Vulnerability. *Global Environmental Change, 16*(3), 268–281.

Baldwin, A. (2013). Racialisation and the figure of the climate-change migrant. *Environment and Planning A, 45*(6), 1474–1490.

Barbier, B., Yacouba, H., Karambiri, H., Zoromé, M., & Somé, B. (2009). Human vulnerability to climate variability in the Sahel: Farmers' adaptation strategies in northern Burkina Faso. *Environmental Management, 43*(5), 790–803.

Bardsley, D. K., & Hugo, G. J. (2010). Migration and climate change: Examining thresholds of change to guide effective adaptation decision-making. *Population and Environment, 32*(2–3), 238–262.

Bettini, G. (2014). Climate migration as an adaption strategy: De-securitizing climate-induced migration or making the unruly governable? *Critical Studies on Security, 2*(2), 180–195.

Biermann, F., & Boas, I. (2012). Climate change and human migration: Towards a global governance system to protect climate refugees. In J. Scheffran et al. (Eds.), *Climate change, human security and violent conflict* (pp. 291–300). Berlin: Springer.

Biggar, J. C. (1979). The sunning of America: Migration to the Sunbelt. *Population Bulletin, 34*(1), 44.

Black, R., Bennett, S., Thomas, S., & Beddington, J. (2011). Migration as adaptation. *Nature, 478*, 447–449.

Black, R., Arnell, N., Adger, W. N., Thomas, D., & Geddes, A. (2012). Migration, immobility and displacement outcomes following extreme events. *Environmental Science and Policy*. doi:10.1016/j.envsci.2012.09.001.

Blaikie, P. (1981). Class, land-use and soil erosion. *Development Policy Review, A14*(2), 57–77.

Brown, O., Hammill, A., & McLeman, R. (2007). Climate change as the "new" security threat: Implications for Africa. *International Affairs, 83*(6), 1141–1154.

Burton, I., Kates, R. W., & White, G. F. (1978). *The environment as hazard*. New York: Guilford Press.
Darwin, C. (1859). *The Origin of Species*. Edison: Castle Books (reprint).
El-Hinnawi, E. (1985). *Environmental refugees*. Nairobi: United Nations Environmental Program.
Felli, R., & Castree, N. (2012). Neoliberalising adaptation to environmental change: Foresight or foreclosure? *Environment and Planning A, 44*(1), 1–4.
Foresight: Migration and Global Environmental Change, 2011. Final Project Report, Government Office for Science, London. Available at: http://www.bis.gov.uk/assets/bispartners/foresight/docs/migration/11-1116-migration-and-global-environmental-change.pdf.
Geddes, A., & Jordan, A. (2012). Migration as adaptation? Exploring the scope for coordinating environmental and migration policies in the European Union. *Environment and Planning C, 30*, 1029–1044.
Gemenne, F., & Brücker, P. (2015). From the guiding principles on internal displacement to the Nansen initiative: What the governance of environmental migration can learn from the governance of internal displacement. *International Journal of Refugee Law, 27*(2), 245–263.
Gibb, C., & Ford, J. (2012). Should the United Nations framework convention on climate change recognize climate migrants? *Environmental Research Letters, 7*(4), 045601.
Glantz, M. H. (1991). The use of analogies in forecasting ecological and societal responses to global warming. *Environment, 33*(5), 10–33.
Hewitt, K. (1983). The idea of calamity in a technocratic age. In K. Hewitt (Ed.), *Interpretations of calamity: From the viewpoint of human ecology* (pp. 3–32). Winchester: Unwin & Allen.
Hunter, L. M. (2005). Migration and environmental hazards. *Population and Environment, 26*(4), 273–302.
Huntington, E. (1924). *Civilization and climate* (3rd ed.). New Haven: Yale University Press.
Kiribati Office of the President. (2015). Relocation. http://www.climate.gov.ki/category/action/relocation/
Little, M. A., & Baker, P. T. (1988). Migration and adaptation. In C. G. N. Mascie-Taylor & G. W. Lasker (Eds.), *Biological aspects of human migration* (pp. 167–215). New York: Cambridge University Press.
Maldonado, J. K., Shearer, C., Bronen, R., Peterson, K., & Lazrus, H. (2013). The impact of climate change on tribal communities in the US: Displacement, relocation, and human rights. *Climatic Change, 120*(3), 601–614.
Manes, S. (1982). Pioneers and survivors: Oklahoma's landless farmers. In A. H. Morgan & H. W. Morgan (Eds.), *Oklahoma: New views of the forty-sixth state* (pp. 93–132). Norman: University of Oklahoma Press.
Mayer, B. (2015). The arbitrary project of protecting environmental migrants. In R. McLeman, J. Schade, & T. Faist (Eds.), *Environmental migration and social inequality* (pp. 189–202). Dordrecht: Springer.
McDean, H. C. (1986). Dust bowl historiography. *Great Plains Quarterly, 6*(2), 117–126.
McLeman, R. (2006). Migration out of 1930s rural eastern Oklahoma: Insights for climate change research. *Great Plains Quarterly, 26*(1), 27–40.
McLeman, R. (2014). *Climate and human migration: Past experiences, future challenges*. New York: Cambridge University Press.
McLeman, R., & Smit, B. (2003). *Climate change, migration and security* (Commentary no. 86). Ottawa: Canadian Security Intelligence Service.
McLeman, R., & Smit, B. (2006). Migration as an adaptation to climate change. *Climatic Change, 76*(1–2), 31–53.
McLeman, R., Faist, T., & Schade, J. (2015). Environment, migration, and inequality – A complex dynamic. In R. McLeman, J. Schade, & T. Faist (Eds.), *Environmental migration and social inequality* (pp. 3–26). Dordrecht: Springer.
Mezdour, A., Veronis, L., & McLeman, R. (2015). Environmental influences on Haitian migration to Canada and connections to social inequality: Evidence from Ottawa-Gatineau and Montreal.

In R. McLeman, J. Schade, & T. Faist (Eds.), *Environmental migration and social inequality* (pp. 103–116). Dordrecht: Springer.

Mortreux, C., & Barnett, J. (2009). Climate change, migration and adaptation in Funafuti, Tuvalu. *Global Environmental Change, 19*(1), 105–112.

Mullins, P. (1979). Australia's sunbelt migration: The recent growth of Brisbane and the Moreton Bay region. *Journal of Australian Political Economy, 5*, 17–32.

Obokata, R., Veronis, L., & McLeman, R. (2014). Empirical research on international environmental migration: A systematic review. *Population and Environment, 36*(1), 111–135.

Perch-Nielsen, S., Bättig, M., & Imboden, D. (2008). Exploring the link between climate change and migration. *Climatic Change, 91*(3–4), 375–393.

Petersen, W. (1958). A general typology of migration. *American Sociological Review, 23*(3), 256–266.

Price, J. (1968). The migration and adaptation of American Indians to Los Angeles. *Human Organization, 27*(2), 168–175.

Ratzel, F. (1902). *Die Erde und Das Leben*. Bibliographisches Institut.

Ravenstein, E. G. (1889). The laws of migration (Second Paper). *Journal of the Royal Statistical Society, 52*(2), 241–305.

Rosenzweig, C., & Hillel, D. (1993). The dust bowl of the 1930s: Analog of greenhouse effect in the great plains? *Journal of Environmental Quality, 22*(1), 9–22.

Scheffran, J., Marmer, E., & Sow, P. (2012). Migration as a contribution to resilience and innovation in climate adaptation: Social networks and co-development in Northwest Africa. *Applied Geography, 33*, 119–127.

Semple, E. C. (1911). *Influences of geographic environment on the Basis of Ratzel's system of anthropo-geography*. New York: Henry Holt and Company.

Sen, A. (1977). Starvation and exchange entitlements: A general approach and its application to the great Bengal famine. *Cambridge Journal of Economics, 1*(1), 33–59.

Spencer, H. (1860, January). The social organism. *The Westminster Review*.

Svart, L. M. (1976). Environmental preference migration: A review. *Geographical Review, 66*(3), 314–330.

Tacoli, C. (2009). Crisis or adaptation? Migration and climate change in a context of high mobility. *Environment and Urbanization, 21*(2), 513–525.

Veronis, L., & McLeman, R. (2014). Environmental influences on African migration to Canada: Focus group findings from Ottawa-Gatineau. *Population and Environment, 36*(2), 234–251.

Warner, K., & Afifi, T. (2014). Where the rain falls: Evidence from 8 countries on how vulnerable households use migration to manage the risk of rainfall variability and food insecurity. *Climate and Development, 6*(1), 1–17.

Warner, K., Kälin, W., Martin, S., & Nassef, Y. (2015). National adaptation plans and human mobility. *Forced Migration Review, 49*, 8–9.

Watkins, A. J. (1978). Intermetropolitan migration and the rise of the Sunbelt. *Social Science Quarterly, 59*(3), 553–561.

Wolpert, J. (1966). Migration as an adjustment to environmental stress. *Journal of Social Issues, 22*(4), 92–102.

Wyett, K. (2014). Escaping a rising tide: Sea level rise and migration in Kiribati. *Asia & the Pacific Policy Studies, 1*(1), 171–185.

Printed by Printforce, the Netherlands